TRENDS IN GENERAL RELATIVITY AND QUANTUM COSMOLOGY

Trends in General Relativity and Quantum Cosmology

Charles V. Benton
Editor

Nova Science Publishers, Inc.
New York

Copyright © 2006 by Nova Science Publishers, Inc.

All rights reserved. No part of this book may be reproduced, stored in a retrieval system or transmitted in any form or by any means: electronic, electrostatic, magnetic, tape, mechanical photocopying, recording or otherwise without the written permission of the Publisher.

For permission to use material from this book please contact us:
Telephone 631-231-7269; Fax 631-231-8175
Web Site: http://www.novapublishers.com

NOTICE TO THE READER

The Publisher has taken reasonable care in the preparation of this book, but makes no expressed or implied warranty of any kind and assumes no responsibility for any errors or omissions. No liability is assumed for incidental or consequential damages in connection with or arising out of information contained in this book. The Publisher shall not be liable for any special, consequential, or exemplary damages resulting, in whole or in part, from the readers' use of, or reliance upon, this material.

This publication is designed to provide accurate and authoritative information with regard to the subject matter covered herein. It is sold with the clear understanding that the Publisher is not engaged in rendering legal or any other professional services. If legal or any other expert assistance is required, the services of a competent person should be sought. FROM A DECLARATION OF PARTICIPANTS JOINTLY ADOPTED BY A COMMITTEE OF THE AMERICAN BAR ASSOCIATION AND A COMMITTEE OF PUBLISHERS.

LIBRARY OF CONGRESS CATALOGING-IN-PUBLICATION DATA
Trends in general relativity and quantum cosmology / Charles V. Benton (editor).
    p. cm.
Includes bibliographical references and index.
ISBN 1-59454-794-7
1. General relativity (Physics)--Research. 2. Quantum cosmology--Research. I. Benton, Charles V.
QC173.6.T74   2006
530.11--dc22   2005028684

Published by Nova Science Publishers, Inc. ◘*New York*

CONTENTS

Preface vii

Chapter 1 Classical and Quantum Cosmology of an Accelerating Model Universe with Compactification of Extra Dimensions
F. Darabi — 1

Chapter 2 The Fermi Paradox in the Light of the Inflationary
Beatriz Gato-Rivera — 27

Chapter 3 On *3+1* Dimensional Scalar Field Cosmologies
F.L. Williams, P.G. Kevrekidis, T. Christodoulakis, C. Helias, G.O. Papadopoulos and Th. Grammenos — 37

Chapter 4 Notes on Dilaton Quantum Cosmology
Gabriel Catren and Claudio Simeone — 49

Chapter 5 Quantum Mechanics Emerging from "Timeless" Classical Dynamics
Hans-Thomas Elze — 79

Chapter 6 Canonical Structure of 3D Gravity with Torsion
M. Blagojević and B. Cvetković — 103

Chapter 7 Cosmological Pressure Fluctuations and Spatial Expansion
Dale R. Koehler — 125

Index 151

PREFACE

Cosmology deals with the nature of the universe. It can be broadly divided into three great ages. The first began in the 6th century BC with the Pythagorean concept of a spherical Earth that is part of a universe in which the motions of the planets are governed by the harmonious relations of natural laws. The second began in the 16th century with the Copernican revolution. This in turn led into Newton's infinite universe. The third began in the early 20th century with Albert Einstein's theory of general relativity and developed into the expanding universe we know today. Einstein's general theory of relativity extended the new space and time concepts of the special theory of relativity from the domain of electric and magnetic phenomena to all of physics and, particularly, to the theory of gravitation. By building on Einstein's previous work on special relativity, general relativity sought to deal with accelerating frames of reference. This in turn led to the principle of equivalence. By dealing with accelerating frames of reference, general relativity provides astronomers with the best theory to predict the effects of gravity. The new book examines in detail new and important work in this field.

Chapter 1 studies a $(4+D)$-dimensional Kaluza-Klein cosmology with a Robertson-Walker type metric having two scale factors a and R, corresponding to D-dimensional internal space and 4-dimensional universe, respectively. By introducing an exotic matter in the form of a perfect fluid with an special equation of state, as the space-time part of the higher dimensional energy-momentum tensor, a four dimensional effective decaying cosmological term appears as $\lambda \sim R^{-m}$ with $0 \leq m \leq 2$, playing the role of an evolving dark energy in the universe. By taking $m=2$, which has some interesting implications in reconciling observations with inflationary models and is consistent with quantum tunneling, the resulting Einstein's field equations yield the exponential solutions for the scale factors a and R. These exponential behaviors may account for the dynamical compactification of extra dimensions and the accelerating expansion of the 4-dimensional universe in terms of Hubble parameter, H. The acceleration of the universe may be explained by the negative pressure of the exotic matter. It is shown that the rate of compactification of higher dimensions as well as expansion of 4-dimensional universe depends on the dimension, D. The authors then obtain the corresponding Wheeler-DeWitt equation and find the general exact solutions in D-dimensions. A good correspondence between the solutions of classical Einstein's equations and the solutions of the quantum Wheeler-DeWitt equation in any dimension, D, is obtained based on Hartle's point of view concerning the classical limits of quantum cosmology.

The Fermi Paradox is discussed, in chapter 2, in the light of the inflationary and brane world cosmologies. The authors conclude that some brane world cosmologies may be of relevance for the problem of the civilizations spreading throughout our galaxy, but not the inflationary cosmologies, as has been proposed recently. The reason is that cosmological inflation, even if it produces a very old or infinite Universe, like in eternal inflation models, still has little or no influence on the age of our galaxy and is only relevant at much larger scales, which are far beyond visitation or colonization by technological civilizations. Brane world cosmologies, however, have the potential to strengthen the Fermi Paradox. The reason is that in the brane world scenarios our observable Universe is located in a subspace embedded in a much larger spacetime with, at least, one more extra spatial dimension. Along the large extra spatial dimensions there may be other universes. If some of them had the same laws of physics as ours, one can speculate about advanced civilizations able to travel through extra dimensions for visitation or colonization purposes, in either direction.

In chapter 3, the authors analyze the case of *3+1* dimensional scalar field cosmologies in the presence, as well as in the absence of spatial curvature, in isotropic, as well as in anisotropic settings. Our results extend those of Hawkins and Lidsey, by including the non-flat case. The Ermakov-Pinney methodology is developed in a general form, allowing through the converse results presented herein to use it as a tool for constructing new solutions to the original equations. As an example of this type a special blowup solution recently obtained in Christodoulakis *et al.* is retrieved. Additional solutions of the 3+1 dimensional gravity coupled with the scalar field are also obtained. To illustrate the generality of the approach, they extend it to the anisotropic case of Bianchi types I and V and present some related open problems.

In chapter 4, the authors address the canonical quantization of the cosmological models which appear as solutions of the low energy effective action of closed bosonic string theory (dilaton models). The analysis will be restricted to the quantization of the minisuperspace models given by homogeneous and isotropic cosmological solutions. They shall study the different conceptual and technical problems arising in the Hamiltonian formulation of these models as a consequence of the presence of so called Hamiltonian constraint. In particular they shall address the problem of time in quantum cosmology, the characterization of the symmetry under clock reversals arising from the existence of a Hamiltonian constraint and the problem of imposing boundary conditions on the space of solution of the Wheeler–DeWitt equation.

Chapter 5 discusses classical Hamiltonian systems in which the intrinsic proper time evolution parameter is related through a probability distribution to the physical time, which is assumed to be discrete. In this way, a physical clock with discrete states is introduced, which presently is still treated as decoupled from the system. This is motivated by the recent discussion of 'timeless' reparametrization invariant models, where discrete physical time has been constructed based on quasi-local observables. Employing the path-integral formulation of classical mechanics developed by Gozzi et al., the authors show that these deterministic classical systems can be naturally described as unitary quantum mechanical models. They derive the emergent quantum Hamiltonian in terms of the underlying classical one. Such Hamiltonians typically need a regularization — here performed by discretization — in order to arrive at models with a stable groundstate in the continuum limit. This is demonstrated in several examples, recovering and generalizing a model advanced by 't Hooft.

Chapter 6 reports on the canonical structure of the topological 3D gravity with torsion, assuming the anti-de Sitter asymptotic conditions. It is shown that the Poisson bracket algebra of the canonical generators has the form of two independent Virasoro algebras with classical central charges. In contrast to the case of general relativity with a cosmological constant, the values of the central charges are different from each other.

Most recently, experimental determinations of the spectrometric characteristics and internal structural velocities of galaxies have suggested the presence of massive central black holes. The analyses of the galactic spectrometric electromagnetic frequency shifts have resulted in a correlation between the hole mass and the surrounding bulge mass. In chapter 7, the authors examine whether conditions existed in the early universe, that could have led to the formation of gravitational structures possessing such unusual characteristics. They propose an early time pressure fluctuation model, which would have generated a radiation based energy distribution possessing the characteristic of a centrally collapsed zone isolated from its surrounding environment and thereby manifesting such a black hole behavior. As a hole-core expansion model, it exhibits a time evolving matter and radiation distribution, leading to a supplementary treatment of early time cosmological energy fluctuations. Einstein's gravitational equations are assumed to apply within the radiation-dominated hole-core spatial domain and, with utilization of a spherically symmetric isotropic metric, are used in order to calculate the evolutionary time expansion characteristics. Birth times for the radiation structures are uniquely correlated with the size of the spheres and are primarily determined from the early time energy densities and the apparent curvatures presented by the gravitational equations. Pressure and temperature characteristics are calculated. The hole-core model is described as a flat metric, matter plus radiation, $\sigma = 1/3$, energy distribution. It displays an early time pressure fluctuation collapse, tentatively interpreted to be the formation of a galaxy hole, and therein provides a theoretical basis for the experimental data.

Chapter 1

CLASSICAL AND QUANTUM COSMOLOGY OF AN ACCELERATING MODEL UNIVERSE WITH COMPACTIFICATION OF EXTRA DIMENSIONS

F. Darabi[1]

Department of Physics, Azarbaijan University of Tarbiat Moallem,
53714-161, Tabriz, Iran.
Institute for Studies in Theoretical Physics and Mathematics, Farmanieh,
19395-5531, Tehran, Iran.

ABSTRACT

We study a $(4+D)$-dimensional Kaluza-Klein cosmology with a Robertson-Walker type metric having two scale factors a and R, corresponding to D-dimensional internal space and 4-dimensional universe, respectively. By introducing an exotic matter in the form of a perfect fluid with an special equation of state, as the space-time part of the higher dimensional energy-momentum tensor, a four dimensional effective decaying cosmological term appears as $\lambda \sim R^{-m}$ with $0 \leq m \leq 2$, playing the role of an evolving dark energy in the universe. By taking $m=2$, which has some interesting implications in reconciling observations with inflationary models and is consistent with quantum tunneling, the resulting Einstein's field equations yield the exponential solutions for the scale factors a and R. These exponential behaviors may account for the dynamical compactification of extra dimensions and the accelerating expansion of the 4-dimensional universe in terms of Hubble parameter, H. The acceleration of the universe may be explained by the negative pressure of the exotic matter. It is shown that the rate of compactification of higher dimensions as well as expansion of 4-dimensional universe depends on the dimension, D. We then obtain the corresponding Wheeler-DeWitt equation and find the general exact solutions in D-dimensions. A good correspondence between the solutions of classical Einstein's equations and the solutions of the quantum Wheeler-DeWitt equation in any dimension, D, is obtained based on Hartle's point of view concerning the classical limits of quantum cosmology.

---

[1] E-mail: f.darabi@azaruniv.edu

INTRODUCTION

Cosmological models with a cosmological term Λ are currently serious candidates to describe the dynamics of our four dimensional universe. The history of cosmological term dates back to Einstein, and its original role was to allow static homogeneous solutions to Einstein's equations in the presence of matter which turned out to be unnecessary when the expansion of the universe was discovered. However, particle physicists then realized that the non-vanishing cosmological constant can be interpreted as a measure of the energy density of the vacuum which turned out to be the sum of a number of apparently disjoint contributions of quantum fields. In fact, a dynamical characteristic for the vacuum energy density (cosmological term) was attributed by quantum field theorists since the developments in particle physics and inflationary scenarios. According to modern quantum field theory, the structure of a vacuum turns out to be interrelated with some spontaneous symmetry-breaking effects through the condensation of quantum (scalar) fields. This phenomenon gives rise to a non-vanishing vacuum energy density of the form $<T_{\mu\nu}> = -<\rho> g_{\mu\nu}$. Therefore, the observed (or effective) cosmological term receives an extra contribution from $<T_{\mu\nu}>$ as follows:

$$\Lambda = \lambda + 8\pi G <\rho>,$$

where λ is the bare cosmological constant and G is the gravitational constant. From quantum field theory we may expect $<\rho> \approx M_{Pl}^4 \approx 2 \times 10^{71} GeV^4$ (M_{Pl} is the Planck mass), or another energy scale related to some spontaneous symmetry breaking effect such as M_{SUSY}^4 or M_{Weak}^4. Therefore, the bare cosmological constant receives potential contributions from these mass scales resulting in a large effective cosmological term. However, the experimental upper bound on the present value of the cosmological term, Λ, provided by measurements of the Hubble constant, H, reads numerically as

$$\frac{|\Lambda|}{8\pi G} \leq 10^{-29} g/cm^3 \approx 10^{-47} GeV^4,$$

which is too far from the expectation of quantum field theory. The question of why the observed vacuum energy is so small in comparison to the scales of particle physics is known as the cosmological constant problem. It is generally thought to be easier to imagine an unknown mechanism which would set Λ exactly to zero than one which would suppress it by just the right amount to yield an observationally tiny cosmological constant. If Λ is a dynamical variable (or vacuum parameter), then it is natural to suppose that in an expanding universe the cosmological term relaxes to the present tiny value by some relaxation mechanism which may be provided by a time-varying vacuum with a rolling scalar field [2].

There are still other possibilities to be advocated. In recent years, several attempts in these directions have been done, in the context of quantum cosmology [3]. One plausible explanation for a tiny cosmological term is to suppose that Λ is dynamically evolving and not constant, i.e., $\Lambda \propto R^{-m}$, where R is the scale factor of the universe and m is a parameter. So, as the universe expands from its small size in the early universe, the initially large effective cosmological term evolves and reduces to its present small value[4].

The study of Λ-decaying cosmological models has recently been the subject of particular interest both from classical and quantum aspects. The Λ decaying models may serve as potential candidates to solve this problem by decaying the large value of the cosmological constant Λ to its present observed value.

Also, there are strong (astronomical) observational motivations for considering cosmological models in which Λ is dynamically decreasing as $\Lambda \propto R^{-m}$. Some models assume *a priori* a fixed value for the parameter m. The case $m=2$, corresponding to the cosmic string matter, has mostly been taken based on dimensional considerations by some authors [5]. The case $m \approx 4$ which resembles the ordinary radiation has also been considered by some other authors [6]. A third group of authors have also studied the case $m=3$ corresponding to the ordinary matter [7]. There are also some other models in which the value of m is not fixed *a priori* and the numerical bounds on the value of m is estimated by observational data or obtained by calculation of the quantum tunnelling rate [8]. Other aspects of Λ-decaying models have also been discussed with no specific numerical bounds on m [9]. It is clear that the functional dependence $\Lambda \propto R^{-m}$ is phenomenological and does not result from the first principles of particle physics. However, for some domain for example, $0 \leq m < 3$, the decaying law $\Lambda \propto R^{-m}$ deserves further investigation. One important reason is that the age of the universe, in these models, is always larger than the age obtained in the standard Einstein-de Sitter cosmology, or the one we get in an open universe. Therefore, if we are interested in solving the age problem, the decaying Λ term appears to be a good candidate. In fact, according to the ansatz $\Lambda \propto R^{-m}$, one may suppose the natural value $<\rho> \approx M_{Pl}^4$ to be the value of Λ at the Planck time when R was of the order of the Planck length. Theoretically this ansatz does not directly solve the cosmological constant problem, but it relates this problem to the age problem of why our universe is so old and have a radius R much larger than the Planck length. In other words, this ansatz reduces two above problems to one problem of "Why our universe could have escaped the death at the Planck time", which seems to be the most natural fate of a baby-universe in quantum cosmology? One may assume that the value of Λ in the early universe might have been much bigger than its present value and large enough to drive some symmetry breakings which might have occurred in the early universe.

On the other hand, the idea that our 4-dimensional universe might have emerged from a higher dimensional space-time is now receiving much attention [10] where the compactification of higher dimensions plays a key role. However, the question of how and why this compactification occurs remains as an open problem. From string theory we know that the compactification may take place provided that the higher dimensional manifold admits special properties, namely if the geometry of the manifold allows, for example, the existence of suitable Killing vectors. However, it is difficult to understand why such manifolds are preferred and whether other possible mechanisms for compactification do exist. In cosmology, on the other hand, different kinds of compactifications could be considered. For example, in an approach, called *dynamical compactification*, the extra dimensions evolve in time towards very small sizes and the extra-dimensional universe reduces to an effective four-dimensional one. This type of compactification was considered in the context of Modern Kaluza-Klein theories [18]. It is then a natural question that how an effective four dimensional universe evolve in time and whether the resulting cosmology is similar to the standard Friedmann-Robertson-Walker four dimensional universe without extra dimensions.

Meanwhile, the recent distance measurements of type Ia *supernova* suggest strongly an accelerating universe [11]. This accelerating expansion is generally believed to be driven by an energy source called *dark energy* which provides negative pressure, such as a positive cosmological constant [12], or a slowly evolving real scalar field called *quintessence* [13]. Moreover, the basic conclusion from all previous observations that ~ 70 percent of the energy density of the universe is in a dark energy sector, has been confirmed after the recent WMAP [14].

To model a universe based on these considerations one may start from a fundamental theory including both gravity and standard model of particle physics. In this regards, it is interesting to begin with ten or eleven-dimensional space-time of superstring/M-theory, in which case one needs a compactification of ten or eleven-dimensional supergravity theory where an effective 4-dimensional cosmology undergoes acceleration. However, it has been known for some time that it is difficult to derive such a cosmology and has been considered that there is a no-go theorem that excludes such a possibility, if one takes the internal space to be time-independent and compact without boundary [15]. However, it has recently been shown that one may avoid this no-go theorem by giving up the condition of time-independence of the internal space; and a solution of the vacuum Einstein equations with compact hyperbolic internal space has been proposed based on this model [16]. Similar accelerating cosmologies can also be obtained for SM2 and SD2 branes, not only for hyperbolic but also for flat internal space [17].

On the other hand, from cosmological point of view, it is not so difficult to find cosmological models in which the 4-dimensional universe undergoes an accelerating expansion and the internal space contracts with time, exhibiting the *dynamical compactification* [18], [19], [20].

In [20], for instance, it is shown that using a more general metric, as compared to Ref.[18], and introducing matter without specifying its nature, the size of compact space evolves as an inverse power of the radius of the universe. The Friedmann-Robertson-Walker equations of the standard four-dimensional cosmology is obtained using an effective pressure expressed in terms of the components of the higher dimensional energy-momentum tensor, and the negative value of this pressure may explain the acceleration of our present universe.

To the author's knowledge the question of Λ-decaying cosmological model has not received much attention in higher dimensional Kaluza-Klein cosmologies. Moreover, the exotic matter has not been considered as an alternative candidate to produce the acceleration of the universe. The purpose of the present chapter is to study a $(4+D)$-dimensional Kaluza-Klein cosmology, with an extended Robertson-Walker type metric, in this context [1]. As we are concerned with cosmological solutions, which are intrinsically time dependent, we may suppose that the internal space is also time dependent. It is shown that by taking this higher dimensional metric and introducing a 4-dimensional exotic matter, a decaying cosmological term $\Lambda \sim R^{-m}$ with $0 \leq m \leq 2$ is appeared as a type of dark energy, and for the case $m = 2$ the resulting field equations yield the exponential solutions for the scale factors of the four-dimensional universe and the internal space. These solutions may account for the accelerating universe and dynamical compactification of extra dimensions, driven by the negative pressure of the exotic matter [2]. It should be noted, however, that the solutions in principle describe typical inflation rather than the recently observed acceleration of the universe which is known to take place in an ordinary matter dominated universe. Nevertheless, regarding the fact that about 70 percent of the total energy density of the universe is of dark energy type with negative pressure, we may approximate the matter content of the universe with almost dark energy and consider the present model as a rather simplified model of a real accelerating universe.

The quantum cosmology of this model is also studied by obtaining the Wheeler-DeWitt equation and finding its general exact solutions. It is then shown that a good correspondence exists between the classical and quantum cosmological solutions, based on the interpretation of Hartle of the classical limits of quantum cosmology.

The chapter is organized as follows: In section, we introduce the classical cosmology model by taking a higher dimensional Robertson-Walker type metric and a higher dimensional matter whose non-zero part is a four-dimensional exotic matter. In section, we obtain the Einstein equations for the two scale factors. In section, we solve the Einstein equations and obtain the solutions. In section, we study the corresponding quantum cosmology and derive the Wheeler-DeWitt equation. In section, the exact solutions of the Wheeler-DeWitt equation is obtained. Finally, in section, we show a good correspondence between the classical and quantum cosmology. The chapter is ended with concluding remarks.

CLASSICAL COSMOLOGY

To begin with, we study the metric considered in [9] in which the space-time is assumed to be of Robertson-Walker type having a (3+1)-dimensional space-time part and an internal space with dimension D. We adopt a real chart $\{t, r^i, \rho^a\}$ with t, r^i, and ρ^a denoting the time, space coordinates and internal space dimensions, respectively. We, therefore, take[3]

$$ds^2 = -N^2(t)dt^2 + R^2(t)\frac{dr^i dr^i}{(1+\frac{kr^2}{4})^2} + a^2(t)\frac{d\rho^a d\rho^a}{(1+k'\rho^2)^2}, \tag{1}$$

where $N(t)$ is the lapse function, $R(t)$ and $a(t)$ are the scale factor of the universe and the radius of internal space, respectively; $r^2 \equiv r^i r^i (i = 1,2,3), \rho^2 \equiv \rho^a \rho^a (a = 1,...D)$, and $k, k' = 0, \pm 1$, reflecting flat, open or closed type of four-dimensional universe and D-dimensional space. We assume the internal space to be flat with compact topology S^D, which means $k' = 0$. This assumption is motivated by the possibility of the compact spaces

---

[2] A similar work [21] has already been done in which the same extended FRW metric was chosen with a radiation fluid occupying all the extended space-time. They found inflation for 3-dimensions and a contraction for the D remaining spatial dimensions.

[3] There is a little difference between this metric and that of [9], in that here the lapse function is generally considered as $N(t)$ instead of taking $N=1$.

to be flat or hyperbolic in "*accelerating cosmologies from compactification*" scenarios, as discussed in Introduction.

The form of energy-momentum tensor is dictated by Einstein's equations and by the symmetries of the metric (1). Therefore, we may assume

$$T_{AB} = (-\rho, p, p, p, p_D, p_D, ..., p_D), \quad (2)$$

where A and B run over both the space-time coordinates and the internal space dimensions. Now, we examine the case for which the pressure along all the extra dimensions vanishes, namely $p_D = 0$. In so doing, we are motivated by the *brane world* scenarios where the matter is to be confined to the 4-dimensional universe, so that all components of T_{AB} is set to zero but the space-time components [23] and it means no matter escapes through the extra dimensions.

We assume the energy-momentum tensor $T_{\mu\nu}$ of space-time to be an exotic χ fluid with the equation of state

$$p_\chi = (\frac{m}{3} - 1)\rho_\chi, \quad (3)$$

where p_χ and ρ_χ are the pressure and density of the fluid, respectively and the parameter m is restricted to the range $0 \leq m \leq 2$ [24]. It is worth noting that the equation of state (3) with $0 \leq m \leq 2$ resembles a universe with negative pressure matter, violating the strong energy condition [25] and this violation is required for a universe to be accelerated [16][4].

Using standard techniques we obtain the scalar curvature corresponding to the metric (1)

$$R = \frac{-6Ra N\ddot{R} + 6Ra\dot{N}\dot{R} - 2R^2 \ddot{a}N + 2R^2 \dot{N}\dot{a} - 2aN^3 k + aN^3 k^2 r^2 - 6aN\dot{R}^2 - 6R\dot{R}\dot{a}N}{R^2 N^3 a},$$

and then substitute it into the dimensionally extended Einstein-Hilbert action (without higher dimensional cosmological term) plus a matter term indicating the above mentioned exotic fluid. This leads to the effective Lagrangian [5]

$$L = \frac{1}{2N} R a^D \dot{R}^2 + \frac{D(D-1)}{12N} R^3 a^{D-2} \dot{a}^2 + \frac{D}{2N} R^2 a^{D-1} \dot{R}\dot{a} - \frac{1}{2} kNRa^D + \frac{1}{6} N\rho_\chi R^3 a^D, \quad (4)$$

where a dot represents differentiation with respect to t. We now take a closed $(k=1)$ universe. Although the flat universe $(k=0)$ is almost favored by observations, we will show an equivalence between $(k=1)$ and $(k=0)$ universes. One may obtain the continuity equation by using the contracted Bianchi identity in $(4+D)$ dimensions, namely

---

[4] Given Einstein equations, this condition on the energy-momentum tensor implies a condition on Ricci tensor as R_{00} 0.
[5] We take the Planck units, $G=c=\hbar=1$

$$\nabla_M G^{MN} = \nabla_M T^{MN} = 0,$$

together with the assumption that the matter is confined to (3+1)-dimensional space-time as

$$T_{ab} = T_{\mu a} = 0,$$

which gives rise to

$$\nabla_\mu T^{\mu\nu} = 0,$$

or

$$\dot{\rho}_\chi R + 3(p_\chi + \rho_\chi)\dot{R} = 0. \tag{5}$$

It is easily shown that substituting the equation of state (3) into the continuity equation (5) leads to the following behavior of the energy density in a closed ($k=1$) Friedmann-Robertson-Walker universe [24]

$$\rho_\chi(R) = \rho_\chi(R_0)\left(\frac{R_0}{R}\right)^m \tag{6}$$

where R_0 is the value of the scale factor at an arbitrary reference time t_0.

Now, if we believe that the cosmological term plays the role of vacuum energy density, we may define the cosmological term [3]

$$\Lambda \equiv \rho_\chi(R), \tag{7}$$

which leads to

$$L = \frac{1}{2N} R a^D \dot{R}^2 + \frac{D(D-1)}{12N} R^3 a^{D-2} \dot{a}^2 + \frac{D}{2N} R^2 a^{D-1} \dot{R}\dot{a} - \frac{1}{2} NRa^D + \frac{1}{6} N\Lambda R^3 a^D, \tag{8}$$

where the cosmological term is now decaying with the scale factor R as

$$\Lambda(R) = \Lambda(R_0)\left(\frac{R_0}{R}\right)^m \tag{9}$$

Note that Λ is now playing the role of an evolving dark energy [26] in 4-dimensions, because we did not consider explicitly a $(4+D)$ dimensional cosmological term in the action, and Λ appears merely due to the specific choice of the equation of state (3) for the exotic matter. The decaying Λ term may also explain the smallness of the present value of the cosmological constant since as the universe evolves from its small to large sizes the large initial value of Λ decays to small values. This phenomenon may somehow alleviate the cosmological constant problem.

Of particular interest, to us, among the different values of m is $m = 2$ which has some interesting implications in reconciling observations with inflationary models [27], and is consistent with quantum tunnelling [3].

EINSTEIN EQUATIONS

We take $m = 2$ and set the initial values of R_0 and $\Lambda(R_0)$ as

$$\Lambda(R_0)R_0^2 = 3 \quad , \quad \Lambda(R) = \frac{3}{R^2}, \tag{10}$$

leading to a positive cosmological term which, according to (7), guarantees the weak energy condition $\rho_\chi > 0$.

The lapse function $N(t)$, in principle, is also an arbitrary function of time due to the fact that Einstein's general relativity is a reparametrization invariant theory. We, therefore, take the gauge

$$N(t) = R^3(t)a^D(t). \tag{11}$$

Now, the Lagrangian becomes

$$L = \frac{1}{2}\frac{\dot{R}^2}{R^2} + \frac{D(D-1)}{12}\frac{\dot{a}^2}{a^2} + \frac{D}{2}\frac{\dot{R}\dot{a}}{Ra}, \tag{12}$$

where Eq.(10) has been used. It is seen that the parameters k and Λ are effectively removed from the Lagrangian and this implies that although k and Λ are not zero in this model the corresponding 4-dimensional universe is equivalent to a flat universe with a zero cosmological term. In other words, we do not distinguish between our familiar 4-dimensional universe, which seems to be flat and without any exotic fluid, and a closed universe filled with an exotic fluid.

We now define the new variables

$$X = \log R \quad , \quad Y = \log a. \tag{13}$$

The lagrangian (12) is written as

$$L = \frac{1}{2}\dot{X}^2 + \frac{D(D-1)}{12}\dot{Y}^2 + \frac{D}{2}\dot{X}\dot{Y}. \tag{14}$$

The equations of motion are obtained

$$\ddot{X} + \frac{D}{2}\ddot{Y} = 0, \qquad (15)$$

$$\ddot{X} + \frac{D-1}{3}\ddot{Y} = 0. \qquad (16)$$

Combining the equations (15) and (16) we obtain

$$\ddot{X} = 0, \qquad (17)$$

$$\ddot{Y} = 0. \qquad (18)$$

SOLUTIONS OF EINSTEIN EQUATIONS

The solutions for X and Y in Eqs. (17) and (18) are obtained

$$X = At + \gamma, \qquad (19)$$

$$Y = Bt + \delta, \qquad (20)$$

and the solutions for $R(t)$ and $a(t)$ are then as follows

$$R(t) = Ae^{\alpha t}, \qquad (21)$$

$$a(t) = Be^{\beta t}, \qquad (22)$$

where the constants "A, B, γ and δ" or "A, B, α and β" should be obtained, in principle, in terms of the initial conditions. It is a reasonable assumption that the size of all spatial dimensions be the same at $t = 0$. Moreover, it may be assumed that this size would be the Planck size l_p in accordance with quantum cosmological considerations. Therefore, we take $R(0) = a(0) = l_p$ so that $A = B = l_p$, and

$$R(t) = l_p e^{\alpha t}, \qquad (23)$$

$$a(t) = l_p e^{\beta t}. \qquad (24)$$

It is important to note that the constants α, β are not independent, and a relation may be obtained between them. This is done by imposing the zero energy condition $H = 0$ which is the well-known result in cosmology due to the existence of arbitrary laps function $N(t)$ in the theory. The Hamiltonian constraint is obtained through the Legender transformation of the Lagrangian (14)

$$H = \frac{1}{2}\dot{X}^2 + \frac{D(D-1)}{12}\dot{Y}^2 + \frac{D}{2}\dot{X}\dot{Y} = 0, \tag{25}$$

which is written in terms of α and β as

$$H = \frac{1}{2}\alpha^2 + \frac{D(D-1)}{12}\beta^2 + \frac{D}{2}\alpha\beta = 0. \tag{26}$$

This constraint is satisfied only for $\alpha \leq 0, \beta \geq 0$ or $\alpha \geq 0, \beta \leq 0$.

For $D \neq 1$, the case $\alpha = 0$ or $\beta = 0$ gives rise to time independent scale factors, namely $R = a = l_p$, which is not physically viable since we know, at least based on observations, the scale factor of the universe is time dependent. We, therefore, choose $\alpha > 0, \beta < 0$ so that the universe and the internal space would expand and contract, respectively, in accordance with the present observations.

For the case $D = 1$, we find

$$\begin{cases} \beta = \text{arbitrary} \\ \alpha = 0 \end{cases} \text{ or } \alpha = -\beta. \tag{27}$$

The former is not physically viable, since it predicts no time evolution for the universe. The latter, however, may predict exponential expansion for $R(t)$, and exponential contraction for $a(t)$, both with the same exponent $\alpha > 0$.

For the general case $D > 1$, we find

$$\alpha_{\pm} = \frac{D\beta}{2}\left[-1 \pm \sqrt{1 - \frac{2}{3}(1 - \frac{1}{D})}\right], \tag{28}$$

which gives two positive values for α indicating two possible expanding universes provided $\beta < 0$ which indicates the compactification of extra dimensions. Moreover, the values of $\alpha_{\pm}$, for a given negative value of β, become larger for higher dimensions. Therefore, the universe expands more rapidly in both possibilities. On the contrary, for a given positive value of α, indicating an expanding universe, the parameter β may take two negative values

$$\beta_{\pm} = \frac{2\alpha}{D}\left[-1 \pm \sqrt{1 - \frac{2}{3}(1 - \frac{1}{D})}\right]^{-1}, \tag{29}$$

indicating two ways of compactification. Moreover, they become smaller for higher dimensions, exhibiting lower rates of compactification.

To find the constants α, β we first obtain the Hubble parameter for $R(t)$

$$H = \frac{\dot{R}}{R} = \alpha, \qquad (30)$$

by which the constant α is fixed. The observed positive value of H will then justify our previous assumption, $\alpha > 0$. We may, therefore, write the solutions (23) and (24) in terms of the Hubble parameter H as

$$R(t) = l_p e^{Ht}, \qquad (31)$$

$$a(t) = l_p e^{-Ht}, \qquad (32)$$

for $D = 1$, and

$$R(t) = l_p e^{Ht}, \qquad (33)$$

$$a(t)_\pm = l_p e^{\frac{2Ht}{D}\left[-1 \pm \sqrt{1-\frac{2}{3}(1-\frac{1}{D})}\right]^{-1}}, \qquad (34)$$

and

$$R_\pm(t) = l_p e^{\frac{D\beta t}{2}\left[-1 \pm \sqrt{1-\frac{2}{3}(1-\frac{1}{D})}\right]}, \qquad (35)$$

$$a(t) = l_p e^{\beta t}. \qquad (36)$$

for $D > 1$.

For a given $H > 0$, it is seen that the solution corresponding to $D = 1$ may predict an accelerating (de Sitter) universe and a contracting internal space with exactly the same rates. For $D > 1$, in Eqs.(33) and (34), for a given $H > 0$ in the exponent of $R(t)$ the exponent in $a(t)$ takes two negative values and becomes smaller for higher dimensions. This means that while the 4-dimensional (de Sitter) universe is expanding by the rate H, the higher dimensions may be compactified in two possible ways with different rates of compactification as a function of dimension, D. In Eqs.(35) and (36), on the other hand, for a given $\beta < 0$ the exponent in $R(t)$ takes two positive values which become larger for higher dimensions. This also means that while the extra dimensions contract by the rate β, the universe may be expanded in two possible ways with different expansion rates as a function of D.

It is easy to show that the Lagrangian (14) (or the equations of motion) is invariant under the simultaneous transformation

$$R \to R^{-1}, \quad a \to a^{-1}, \qquad (37)$$

which is consistent with the time reversal $t \to -t$. Therefore, four different phases of "*expansion-contraction*" for $R(t)$ and $a(t)$ are distinguished, Eqs.(33) - (36). One may prefer the "*expanding $R(t)$ - contracting $a(t)$*" phase to "*expanding $a(t)$ - contracting $R(t)$*" one, considering the present status of the 4D universe [6].

The deceleration parameter q for the scale factor R is obtained

$$q = -\frac{\ddot{R}R}{\dot{R}^2} = -1. \qquad (38)$$

Observational evidences not only do not rule out the negative deceleration parameter but also puts the limits on the present value of q as $-1 \leq q < 0$ [11]. Therefore, this negative value seems to favor a cosmic acceleration in the expansion of the universe.

In the expansion phase of the closed ($k = 1$) universe the cosmological term Λ decays exponentially with time t as

$$\Lambda(t) = 3 l_p^{-2} e^{-2Ht}, \qquad (39)$$

whereas in the contraction phase ($t \to -t$) it grows exponentially to large values so that at $t = 0$ it becomes extremely large, of the order of M_p^2. This huge value of Λ may be extinguished rapidly by assuming a sufficiently large Hubble parameter H, consistent with the present observations, to alleviate the cosmological constant problem.

QUANTUM COSMOLOGY

An appropriate quantum mechanical description of the universe is likely to be afforded by quantum cosmology which was introduced and developed by DeWitt [28]. In quantum cosmology the universe, as a whole, is treated quantum mechanically and is described by a single wave function, $\Psi(h_{ij}, \phi)$, defined on a manifold (*superspace*) of all possible three geometries and all matter field configurations. The wave function $\Psi(h_{ij}, \phi)$ has no explicit time dependence due to the fact that there is no a real time parameter external to the universe. Therefore, there is no Schrödinger wave equation but the operator version of the Hamiltonian constraint of the Dirac canonical quantization procedure [29], namely vanishing of the variation of the Einstein-Hilbert action S with respect to the arbitrary lapse function N

$$H = \frac{\delta S}{\delta N} = 0,$$

---

[6] For the special case $D=3$, both the Lagrangian (14) and the Hamiltonian constraint (25) are invariant under the transformation $a \to R$, $R \to a$. Therefore, we have a dynamical symmetry between R and a, namely $a \leftrightarrow R$. In this case there is no real line of demarcation between a and R to single out one of them as the real scale factor of the universe. This is because the internal space is flat $k'=0$ and according to (12) one may assume the 4D universe with k, $\Lambda \neq 0$ to be equivalent to the one in which $k=\Lambda=0$. Therefore, both have the same topology S^3.

which is written
$$\hat{H}\Psi(h_{ij},\phi) = 0.$$

This equation is known as the Wheeler-DeWitt (WDW) equation. The goal of quantum cosmology by solving the WDW equation is to understand the origin and evolution of the universe, quantum mechanically. As a differential equation, the WDW equation has an infinite number of solutions. To get a unique viable solution, we should also respect the question of boundary condition in quantum cosmology which is of prime importance in obtaining the relevant solutions for the WDW equation.

In principle, it is very difficult to solve the WDW equation in the *superspace* due to the large number of degrees of freedom. In practice, one has to *freeze out* of all but a finite number of degrees of freedom of the gravitational and matter fields. This procedure is known as quantization in *minisuperspace*, and will be used in the following discussion.

The minisuperspace in our model is two-dimensional with gravitational variables X and Y. To obtain the Wheeler-DeWitt equation, in this minisuperspace, we start with the Lagrangian (14). The conjugate momenta corresponding to X and Y are obtained

$$P_X = \frac{\partial L}{\partial \dot{X}} = \dot{X} + \frac{D}{2}\dot{Y}, \tag{40}$$

$$P_Y = \frac{\partial L}{\partial \dot{Y}} = \frac{D}{2}\dot{X} + \frac{D(D-1)}{6}\dot{Y}, \tag{41}$$

from which we obtain

$$\dot{X} = \frac{6}{D+2}\left[P_X\left(\frac{1-D}{3}\right) + P_Y\right], \tag{42}$$

$$\dot{Y} = \frac{6}{D(D-1)}\left[P_Y\frac{2(1-D)}{D+2} - P_X\frac{D(1-D)}{D+2}\right]. \tag{43}$$

Substituting Eqs.(42), (43) into the Hamiltonian constraint (25), we obtain

$$H = (1-D)P_X^2 - \frac{6}{D}P_Y^2 + 6P_X P_Y = 0. \tag{44}$$

Now, we may use the following quantum mechanical replacements

$$P_X \to -i\frac{\partial}{\partial X}, \quad P_Y \to -i\frac{\partial}{\partial Y},$$

by which the Wheeler-DeWitt equation is obtained

$$\left[(D-1)\frac{\partial^2}{\partial X^2}+\frac{6}{D}\frac{\partial^2}{\partial Y^2}-6\frac{\partial}{\partial X}\frac{\partial}{\partial Y}\right]\Psi(X,Y)=0, \tag{45}$$

where $\Psi(X,Y)$ is the wave function of the universe in the (X,Y) mini-superspace.

We introduce the following change of variables

$$x=X(1-\frac{D}{D+3})+\frac{D}{D+3}Y \quad, \quad y=\frac{X-Y}{D+3}, \tag{46}$$

by which the Wheeler-DeWitt equation takes a simple form

$$\left\{-3\frac{\partial^2}{\partial x^2}+\frac{D+2}{D}\frac{\partial^2}{\partial y^2}\right\}\Psi(x,y)=0. \tag{47}$$

Now, we can separate the variables as $\Psi(x,y)=\phi(x)\psi(y)$ to obtain the following equations

$$\frac{\partial^2\phi(x)}{\partial x^2}=\frac{\gamma}{3}\phi(x), \tag{48}$$

$$\frac{\partial^2\psi(y)}{\partial y^2}=\frac{\gamma D}{D+2}\psi(y), \tag{49}$$

where we assume $\gamma > 0$.

SOLUTIONS OF THE WHEELER-DEWITT EQUATION

The solutions of Eqs.(48), (49) in terms of x, y are as follows

$$\phi(x)=e^{\pm\sqrt{\frac{\gamma}{3}}x}, \tag{50}$$

$$\psi(y)=e^{\pm\sqrt{\frac{\gamma D}{D+2}}y}, \tag{51}$$

leading to the four possible solutions for $\Psi(x,y)$ as

$$\Psi_D^\pm(x,y)=A^\pm e^{\pm\sqrt{\frac{\gamma}{3}}x\pm\sqrt{\frac{\gamma D}{D+2}}y}, \tag{52}$$

$$\Psi_D^\pm(x,y)=B^\pm e^{\pm\sqrt{\frac{\gamma}{3}}x\mp\sqrt{\frac{\gamma D}{D+2}}y}, \tag{53}$$

or alternative solutions in terms of X, Y as

$$\Psi_D^\pm(x,y) = A^\pm e^{\pm\sqrt{\frac{\gamma}{3}}\left(\frac{3X+DY}{D+3}\right)\pm\sqrt{\frac{\gamma D}{D+2}}\left(\frac{X-Y}{D+3}\right)}, \qquad (54)$$

$$\Psi_D^\pm(x,y) = B^\pm e^{\pm\sqrt{\frac{\gamma}{3}}\left(\frac{3X+DY}{D+3}\right)\mp\sqrt{\frac{\gamma D}{D+2}}\left(\frac{X-Y}{D+3}\right)}, \qquad (55)$$

where $A^\pm, B^\pm$ are the normalization constants. We may also write down the solutions in terms of R and a [7]

$$\Psi_D^\pm(R,a) = A^\pm R^{\pm\frac{1}{D+3}\left(\sqrt{3\gamma}+\sqrt{\frac{\gamma D}{D+2}}\right)} a^{\pm\frac{1}{D+3}\left(\sqrt{\frac{\gamma}{3}}D-\sqrt{\frac{\gamma D}{D+2}}\right)}, \qquad (56)$$

$$\Psi_D^\pm(R,a) = B^\pm R^{\pm\frac{1}{D+3}\left(\sqrt{3\gamma}-\sqrt{\frac{\gamma D}{D+2}}\right)} a^{\pm\frac{1}{D+3}\left(\sqrt{\frac{\gamma}{3}}D+\sqrt{\frac{\gamma D}{D+2}}\right)}. \qquad (57)$$

It is now important to impose the *good* boundary conditions on the above solutions to single out the physical ones. In so doing, we may impose the following condition

$$\Psi_D(R\to\infty, a\to\infty) = 0, \qquad (58)$$

which requires the wave function of the universe to be normalizable. This means that our minisuperspace model has no classical solutions that expand simultaneously to infinite values of a and R, as Eqs.(31)-(36) show. Then, one may take the following solutions

$$\Psi_D^\pm(R,a) = C^\pm R^{-\frac{1}{D+3}\left(\sqrt{3\gamma}\pm\sqrt{\frac{\gamma D}{D+2}}\right)} a^{-\frac{1}{D+3}\left(\sqrt{\frac{\gamma}{3}}D\mp\sqrt{\frac{\gamma D}{D+2}}\right)}, \qquad (59)$$

where $C^\pm$ are the normalization constants and the exponents of R and a are negative for any value of D [8].

One may obtain the solutions (59) in (X, Y) mini-superspace as

$$\Psi_D^\pm(x,y) = C^\pm e^{-\sqrt{\frac{\gamma}{3}}\left(\frac{3X+DY}{D+3}\right)\mp\sqrt{\frac{\gamma D}{D+2}}\left(\frac{X-Y}{D+3}\right)}. \qquad (60)$$

---

[7] For $D=3$, there is a exchange symmetry $\Psi(R,a)\leftrightarrow\Psi(a,R)$ under the exchange $a\leftrightarrow R$.

[8] For $D=1$, the exponent of "a" corresponding to Ψ^+ becomes zero so that Ψ^+ depends only on R with the condition $\Psi^+(R\to\infty)\to 0$.

CORRESPONDENCE BETWEEN CLASSICAL AND QUANTUM COSMOLOGY

One of the most interesting topics in the context of quantum cosmology is the mechanisms through which the classical cosmology may emerge from quantum theory. When does a Wheeler-DeWitt wave function predict a classical space-time? Quantum cosmology is the quantum mechanics of an isolated system (universe). It is not possible to use the Copenhagen interpretation, which needs the existence of an external observer, since here the observer is part of the system. Indeed, any attempt in constructing a viable quantum gravity requires understanding the connections between classical and quantum physics. Much work has been done in this direction over the past decade. Actually, there is some tendency towards using semiclassical approximations in dividing the behaviour of the wave function into two types, oscillatory or exponential which are supposed to correspond to classically allowed or forbidden regions. Hartle [30] has put forward a simple rule for applying quantum mechanics to a single system (universe): *If the wave function is sufficiently peaked about some region in the configuration space we predict to observe a correlation between the observables which characterize this region.* Halliwell [31] has shown that the oscillatory semiclassical WKB wave function is peaked about a region of the *minisuperspace* in which the correlation between the coordinate and momentum holds good and stresses that both *correlation* and *decoherence* are necessary before one can say a system is classical. Using Wigner functions, Habib and Laflamme [32] have studied the mutual compatibility of these requirements and shown that some form of coarse graining is necessary for classical prediction from WKB wave functions. Alternatively, Gaussian or coherent states with sharply peaked wave functions are often used to obtain classical limits by constructing wave packets.

In the investigation of classical limits, we first take $D=1$ and look for a correspondence between classical and quantum solutions. Using Eqs.(31) and (32) in the Planck units, the corresponding classical locus in (R,a) configuration space, is

$$Ra = 1, \tag{61}$$

whereas in (X,Y) coordinates we have

$$X + Y = 0. \tag{62}$$

We now consider the wave functions (60) in (X,Y) mini-superspace for $D=1$

$$\Psi_1^+(X,Y) = C^+ e^{-\sqrt{\frac{\gamma}{3}}X}, \tag{63}$$

$$\Psi_1^-(X,Y) = C^- e^{-\sqrt{\frac{\gamma}{3}}\frac{X+Y}{2}}. \tag{64}$$

The above wave functions, in their present form, are not square integrable as is required for the wave functions to predict the classical limit. However, one may take the absolute value of the exponents to make the wave functions square integrable

$$\Psi_1^+(X,Y) = C^+ e^{-|\sqrt{\frac{\gamma}{3}}X|}, \tag{65}$$

$$\Psi_1^-(X,Y) = C^- e^{-|\sqrt{\frac{\gamma}{3}}\frac{X+Y}{2}|}. \tag{66}$$

We next consider the general case $D > 1$. Eliminating the parameter t in Eqs. (33) and (34) the classical loci in terms of R, a are obtained

$$a_\pm = R^{\frac{2}{D}\left[-1\pm\sqrt{1-\frac{2}{3}(1-\frac{1}{D})}\right]^{-1}}. \tag{67}$$

The corresponding forms of these loci in terms of X, Y are

$$Y_+ = \frac{2}{D}X\left[-1+\sqrt{1-\frac{2}{3}(1-\frac{1}{D})}\right]^{-1}, \tag{68}$$

$$Y_- = \frac{2}{D}X\left[-1-\sqrt{1-\frac{2}{3}(1-\frac{1}{D})}\right]^{-1}. \tag{69}$$

The wave functions (60) also are not square integrable, so we may replace the exponents by their absolute values

$$\Psi_D^\pm(x,y) = C^\pm e^{-\left|\sqrt{\frac{\gamma}{3}}\left(\frac{3X+DY}{D+3}\right)\mp\sqrt{\frac{\gamma D}{D+2}}\left(\frac{X-Y}{D+3}\right)\right|}, \tag{70}$$

to make them square integrable. Now, following Hartle's point of view, we try to make correspondence between the classical loci and the wave functions.

Figures 1 - 6 show respectively the 2D plots of the typical wave functions Ψ_1^+ - Ψ_6^+ in terms of (X,Y) for $\gamma = 10^{-6}$; Figures 23 - 28 show the corresponding 3D plots, respectively. On the other hand, Figures 12 - 17 show the classical loci corresponding to $D = 1-6$, respectively. It is seen that the 2D and 3D plots of the wave functions Ψ_1^+ - Ψ_6^+ are exactly peaked on the classical loci.

In the same way, Figures 7 - 11 show respectively the 2D plots of the wave functions Ψ_2^- - Ψ_6^-. Figures 29 - 33 show the corresponding 3D plots, respectively. Figures 18 - 22 show the classical loci for $D = 2-6$, respectively. Again, an exact correspondence is seen between the 2D and 3D plots of the wave functions Ψ_2^- - Ψ_6^- and the classical loci. This procedure will apply for all D.

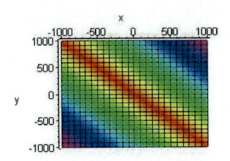

Fig. 1. 2D plot of Ψ_1^+ in terms of (X,Y) for $\gamma = 10^{-6}$

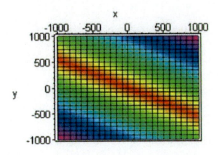

Fig. 2. 2D plot of Ψ_2^+ in terms of (X,Y) for $\gamma = 10^{-6}$

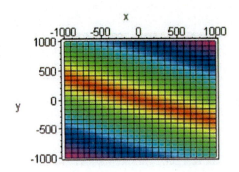

Fig. 3. 2D plot of Ψ_3^+ in terms of (X,Y) for $\gamma = 10^{-6}$

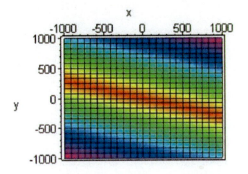

Fig. 4. 2D plot of Ψ_4^+ in terms of (X,Y) for $\gamma = 10^{-6}$

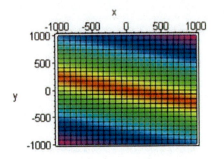

Fig. 5. 2D plot of Ψ_5^+ in terms of (X,Y) for $\gamma = 10^{-6}$

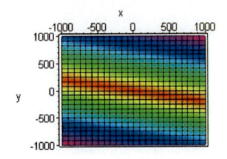

Fig. 6. 2D plot of Ψ_6^+ in terms of (X,Y) for $\gamma = 10^{-6}$

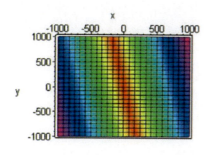

Fig. 7. 2D plot of Ψ_2^- in terms of (X,Y) for $\gamma = 10^{-6}$

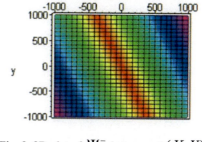

Fig. 8. 2D plot of Ψ_3^- in terms of (X,Y) for $\gamma = 10^{-6}$

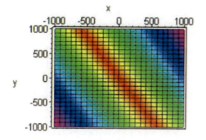

Fig. 9. 2D plot of Ψ_4^- in terms of (X,Y) for $\gamma = 10^{-6}$

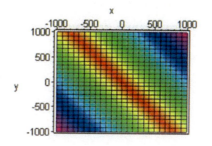

Fig. 10. 2D plot of Ψ_5^- in terms of (X,Y) for $\gamma = 10^{-6}$

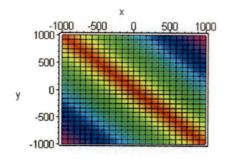

Fig. 11. 2D plot of Ψ_6^- in terms of (X,Y) for $\gamma = 10^{-6}$

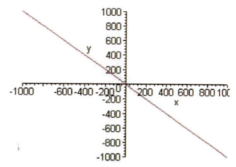

Fig. 12. Classical locus $X + Y = 0$ for $D = 1$

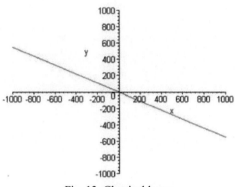

Fig. 13. Classical locus
$Y_- = \frac{2}{D}X[-1-\sqrt{1-\frac{2}{3}(1-\frac{1}{D})}]^{-1}$ for
$D = 2$

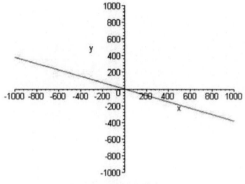

Fig. 14. Classical locus
$Y_- = \frac{2}{D}X[-1-\sqrt{1-\frac{2}{3}(1-\frac{1}{D})}]^{-1}$ for
$D = 3$

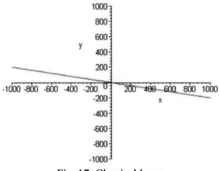

Fig. 15. Classical locus
$Y_- = \frac{2}{D}X[-1-\sqrt{1-\frac{2}{3}(1-\frac{1}{D})}]^{-1}$ for
$D = 4$

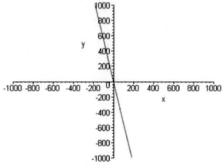

Fig. 16. Classical locus
$Y_- = \frac{2}{D}X[-1-\sqrt{1-\frac{2}{3}(1-\frac{1}{D})}]^{-1}$ for
$D = 5$

Fig. 17. Classical locus
$Y_- = \frac{2}{D}X[-1-\sqrt{1-\frac{2}{3}(1-\frac{1}{D})}]^{-1}$ for
$D = 6$

Fig. 18. Classical locus
$Y_+ = \frac{2}{D}X[-1+\sqrt{1-\frac{2}{3}(1-\frac{1}{D})}]^{-1}$ for
$D = 2$

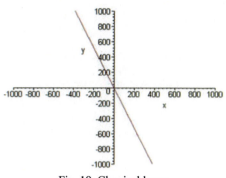

Fig. 19. Classical locus
$Y_+ = \frac{2}{D}X[-1+\sqrt{1-\frac{2}{3}(1-\frac{1}{D})}]^{-1}$ for
$D = 3$

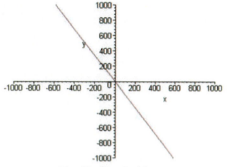

Fig. 20. Classical locus
$Y_+ = \frac{2}{D}X[-1+\sqrt{1-\frac{2}{3}(1-\frac{1}{D})}]^{-1}$ for
$D = 4$

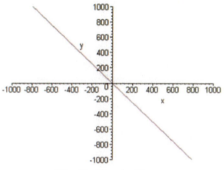

Fig. 21. Classical locus
$Y_+ = \frac{2}{D}X[-1+\sqrt{1-\frac{2}{3}(1-\frac{1}{D})}]^{-1}$ for
$D = 5$

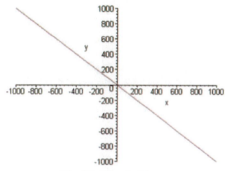

Fig. 22. Classical locus
$Y_+ = \frac{2}{D}X[-1+\sqrt{1-\frac{2}{3}(1-\frac{1}{D})}]^{-1}$ for
$D = 6$

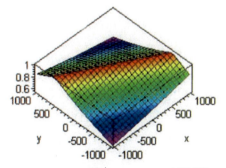

Fig. 23. 3D plot of Ψ_1^+ in terms of (X,Y) for
$\gamma = 10^{-6}$

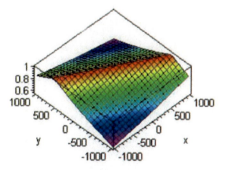

Fig. 24. 3D plot of Ψ_2^+ in terms of (X,Y) for
$\gamma = 10^{-6}$

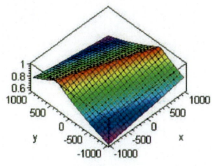

Fig. 25. 3D plot of Ψ_3^+ in terms of (X,Y) for $\gamma = 10^{-6}$

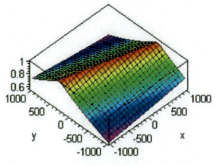

Fig. 26. 3D plot of Ψ_4^+ in terms of (X,Y) for $\gamma = 10^{-6}$

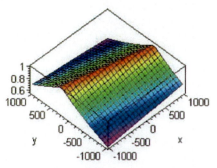

Fig. 27. 3D plot of Ψ_5^+ in terms of (X,Y) for $\gamma = 10^{-6}$

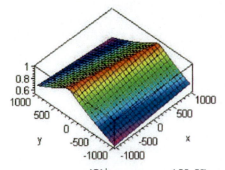

Fig. 28. 3D plot of Ψ_6^+ in terms of (X,Y) for $\gamma = 10^{-6}$

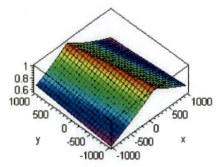

Fig. 29. 3D plot of Ψ_2^- in terms of (X,Y) for $\gamma = 10^{-6}$

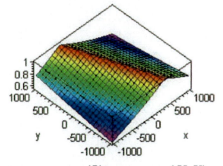

Fig. 30. 3D plot of Ψ_3^- in terms of (X,Y) for $\gamma = 10^{-6}$

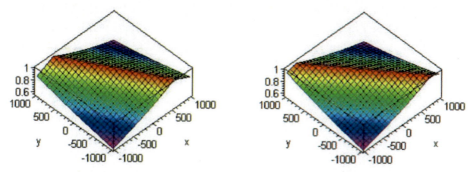

Fig. 31. 3D plot of Ψ_4^- in terms of (X,Y) for $\gamma = 10^{-6}$

Fig. 32. 3D plot of Ψ_5^- in terms of (X,Y) for $\gamma = 10^{-6}$

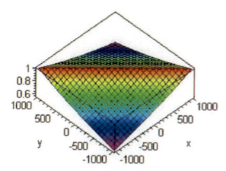

Fig. 33. 3D plot of Ψ_6^- in terms of (X,Y) for $\gamma = 10^{-6}$

CONCLUDING REMARKS

First, we have studied a $(4+D)$-dimensional classical Kaluza-Klein cosmology with a Robertson-Walker type metric having two scale factors, R for the universe and a for the higher dimensional space. By introducing a typical exotic matter with the equation of state $p_\chi = (\frac{m}{3}-1)\rho_\chi$ in 4-dimensions, a decaying cosmological term is obtained effectively as $\lambda \sim R^{-m}$. By taking $m=2$, the corresponding Einstein field equations are obtained and we find exponential solutions for R and a in terms of the Hubble parameter H. These exponential solutions indicate the accelerating expansion of the universe and dynamical compactification of extra dimensions, respectively. It turns out that the rate of compactification of extra dimensions as well as expansion of the universe depends on the number of extra dimensions, D. The more extra dimensions, the less rate of compactification and the more rate of acceleration. It is worth noting that the model is free of initial singularity problem because both R and a are non-zero at $t=0$, resulting in a finite Ricci scalar.

Although the model describes in principle a closed universe with non-vanishing cosmological constant, it is equivalent to a flat universe with zero cosmological constant. Therefore, one may assume that we are really living in a closed universe with $\Lambda \neq 0$, but it effectively appears as a flat universe with $\Lambda = 0$. Note that we have not considered ordinary matter sources in the model except an exotic matter source which is to be considered as a source of dark energy. Therefore, it seems the solutions to describe typical inflation rather than the recently observed acceleration of the universe which is known to take place in an ordinary matter dominated universe. However, if the large percent of the matter sources in the universe would be of dark energy type (as the present observations strongly recommend), then one may keep the results here even in the presence of other matter source, keeping in mind that the relevant contribution to the total matter source of the universe is the dark energy.

A question may arise on the fact that no physics is supposed to exist below the planck length whereas for the contracting solution, the scale factor $a(t)$ goes to zero starting from l_p. However, it is not a major problem because we have not considered elements of quantum gravity theory in this model and merely studied a model based on general relativity which is supposed to be valid in any scale without limitation. The scale l_p, in this paper, is not introduced within a quantum gravity model (action); it just appears as a typical initial condition, in the middle of a classical model, based on the quantum cosmological consideration. One may choose another scale based on some other physical considerations.

We have also studied the corresponding quantum cosmology, through the Wheeler-DeWitt equation, and obtained the exact solutions. Based on Hartle's point of view on the correspondence between the classical and quantum solutions, we have shown by 2D and 3D plots of the wave functions a good correspondence between the classical and quantum cosmological solutions for any D, provided that the wave functions vanish for the infinite scale factors. There is no such a correspondence if another boundary condition, other than stated, is taken. Therefore, this correspondence guaranties that the chosen boundary condition is a *good* one.

REFERENCES

[1] This work is the extended version of the published paper, *Class. Quant. Grav.* 20, 3385 (2003)

[2] S. Adler, *Rev. Mod. Phys.*, 729 (1982); L. F. Abbott, *Phys. Lett.* **B**, 427 (1985); T. Banks, *Nucl. Phys.* **B249**, 332 (1985); S. M. Barr, *Phys. Rev.* D, 1691 (1987); P. J. E. Peebles and B. Ratra, *Astrophys. J. Lett.*, L17 (1988); B. Ratra and P. J. E. Peebles, *Phys. Rev.* D, 3406 (1988).

[3] S. W. Hawking, *Phys. Lett.* **B**, 403 (1984); E. Baum, *ibid.* **B**, 185 (1983); S. Coleman, *Nucl. Phys.* **B307**, 864 (1988); S. BGiddings and A. Strominger, *Nucl. Phys.* **B307**, 854 (1988); **B321**, 481 (1989); T. Banks, *Nucl. Phys.* **B309**, 493 (1988).

[4] M. Ozer and M. O. Taha, *Nucl. Phys.* **B287**, 776 (1987); K. Freese et al, *ibid.* **B287**, 797 (1987); M. Reuter and C. Wetterich, *Phys. Lett.* B, 38 (1987); B. Ratra and P. J. E. Peebles, *Phys. Rev.* D, 3406 (1988); I. Waga, Astrophys. J., 436 (1993); J. A. Frieman

et al., *Phys. Rev. Lett.*, 2077 (1995); K. Coble, S. Dodelson, and J. A. Frieman, *Phys. Rev. D*, 1851 (1997).

[5] J. L. Lopez, D. V. Nanopoulos, Mod. *Phys. Lett.* A, 1 (1996); M. Özer, M. O. Taha, *Phys. Lett.* B, 363 (1986); A.-M. M. Abdel-Rahman, Phys. Rev. D, 3497 (1992); A. and R. G. Vishwakarma, *Class. Quant. Grav.*, 945 (1997); W. Chen, Y.-S. Wu, *Phys. Rev. D*, 695 (1990); M. O. Calvao et al, *Phys. Rev.* D, 3869 (1992); V. Méndez, D. Pavón, Gen. Rel. Grav., 697 (1996).

[6] K. Freese et al, Nucl. Phys. B, 797 (1987); M. Gasperini, *Phys. Lett.* B, 347 (1987); J. M. Overduin, P. S. Wesson and S. Bowyer, *Astrophys. J.*, 1 (1993).

[7] F. Hoyle, G. Burbidge and J. V. Narlikar, *Mon. Not. R. Astron. Soc.*, 173 (1997).

[8] M. A. Jafarizadeh, F. Darabi, A. Rezaei-Aghdam and A. R. Rastegar, *Phys. Rev. D*, 063514 (1999); T. S. Olson, T. F. Jordan, *Phys. Rev.* D, 3258 (1987); V. Silveira, I. Waga, Phys. Rev. D, 4890 (1994); Phys. Rev. D, 4625 (1997); L. F. B. Torres, I. Waga, *Mon. Not. R. Astron. Soc.*, 712 (1996).

[9] M. D. Maia, G. S. Silva, *Phys. Rev.* D, 7233 (1994); R. F. Sisteró, Gen. Rel. Grav., 1265 (1991); J. Matyjasek, *Phys. Rev.* D, 4154 (1995).

[10] T. Appelquist, A. Chodos and P. G. O. Freund, Modern Kaluza-Klein Theories, *Frontiers in Physics Series*, (Volume), 1986, (Ed. Addison-Wesley). A. G. Riess et al. *Astrophys. J.*, 49 (2001).

[11] S. Perlmutter et al., *Bull. Am. Phys. Soc.*, 1351 (1997); Ap. J., 46 (1998); A. G. Riess et al., *Astron. J.* (1998); P. M. Garnavich et al., *Ap. J. Lett.*, 53 (1998); Science, 1298 (1998); *Ap. J.*, 74 (1998); B. Schmidt et al., *Astrophys. J.*, 46 (1998).

[12] L. Krauss and M. S. Turner, *Gen. Rel. Grav.*, 1137 (1995); J. P. Ostriker and P. J. Steinhardt, *Nature*, 600 (1995); A. R. Liddle, D. H. Lyth, P. T. Viana and M. White, Mon. *Not. Roy. Astron. Soc.*, 281 (1996).

[13] R. R. Caldwell, R. Dave and P. J. Steinhardt, *Phys. Rev. Lett.*, 1582 (1998); M. C. Bento, O. Bertolami, *Gen. Rel. Grav.*, 1461 (1999); O. Bertolami, Nuovo Cimento, B, 36 (1986).

[14] C. L. Benett et al., *Astrophys. J. Supll.*, 1 (2003).

[15] G. W. Gibbons, Aspects of Supergravity Theories in Supersymmetry, Supergravity and Related Topics, (World Scientific Singapore, 1985); J. Maldacena, C. Nuñez, *Int. J. Mod. Phys.* A, 822 (2001).

[16] P. K. Townsend, M. N. R. Wohlfarth, *Phys. Rev. Lett.*, 061302 (2003).

[17] L. Cornalba, M. Costa, *Phys. Rev.* D, 066001 (2002); L. Cornalba, M. Costa, and C. Kounnas, Nucl. Phys. B637, 378 (2002); N. Ohta, *Phys. Rev. Lett.*, 061303 (2003); S. Roy, Phys. Lett. **B568**, 1 (2003).

[18] A. Chodos and S. Detweiler, *Phys. Rev.* D, 2167 (1980); T. Dereli and R. W. Tucker, *Phys. Lett.* **B125**, 133 (1983); J. Gu, W. P. Hwang, Phys. Rev. D, 024003 (2002).

[19] A. S. Majumdar and S. K. Sethi, *Phys. Rev.* D, 5315 (1992); A. S. Majumdar, T. R. Seshadri and S. K. Sethi, *Phys. Lett.* B, 67 (1993); A. S. Majumdar, Phys. Rev. D, 6092 (1997); A. S. Majumdar, *Ind. Jour. Phys.* B, 843 (1999); A. S. Majumdar, *Phys. Rev. D*, 083503 (2001).

[20] N. Mohammedi, *Phys. Rev.* D, 104018 (2002).

[21] D. Sahdev, *Phys. Lett.* B 155 (1984).

[22] J. Wudka, *Phys. Rev.* D, 3255 (1987); *Phys. Rev.* D, 1036 (1987).

[23] G. Dvali, G. Gabadadze, M. Porrati, *Phys. Lett.* B, 208 (2000); B. Abdesselam, N. Mohemmedi, *Phys. Rev.* D65, 084018 (2002); N. Kaloper, J. March-Russell, G. D. Starkman and M. Trodden, *Phys. Rev. Lett.*, 928 (2000).
[24] D. Atkatz and H. Pagels, *Phys. Rev.* D, 2065 (1982); Am. J. Phys. (7), 619 (1994).
[25] M. A. Jafarizadeh, F. Darabi, A. Rezaei-Aghdam and A. R. Rastegar, *Mod. Phys. Lett.* A, 3213 (1998).
[26] P. J. E. Peebles, B. Ratra, *Rev. Mod. Phys.*, 559 (2003).
[27] W. Chen and Y. Wu, *Phys. Rev.* D, 695 (1990).
[28] B. S. DeWitt, *Phys. Rev.*, 1113 (1967).
[29] P. M. A. Dirac, *Lectures on Quantum Mechanics*, Yeshiva University, (Academic press, New York) 1967.
[30] J. B. Hartle, in *Gravitation in Astrophysics*, 1986 NATO Advanced Summer Institute, Gargèse, edited by B. Carter and J. Hartle (NATO ASI Series B: Physics Vol., Plenum. New York) 1987.
[31] J. J. Halliwell, *Phys. Rev. D* 2912 (1989).
[32] S. Habib, R. Laflamme, *Phys. Rev. D* 4056 (1990); S. Habib, *Phys. Rev. D* 2566 (1990).

In: Trends in General Relativity and Quantum Cosmology
Editor: Charles V. Benton, pp. 27-35

ISBN 1-59454-794-7
© 2006 Nova Science Publishers, Inc.

Chapter 2

THE FERMI PARADOX IN THE LIGHT OF THE INFLATIONARY

Beatriz Gato-Rivera
Instituto de Matemáticas y Física Fundamental, CSIC
Serrano 123, Madrid 28006, Spain
e-mail address: bgator@imaff.cfmac.csic.es NIKHEF-H, Kruislaan 409,
NL-1098 SJ Amsterdam, The Netherlands

Abstract

The Fermi Paradox is discussed in the light of the inflationary and brane world cosmologies. We conclude that some brane world cosmologies may be of relevance for the problem of the civilizations spreading throughout our galaxy, but not the inflationary cosmologies, as has been proposed recently. The reason is that cosmological inflation, even if it produces a very old or infinite Universe, like in eternal inflation models, still has little or no influence on the age of our galaxy and is only relevant at much larger scales, which are far beyond visitation or colonization by technological civilizations. Brane world cosmologies, however, have the potential to strengthen the Fermi Paradox. The reason is that in the brane world scenarios our observable Universe is located in a subspace embedded in a much larger spacetime with, at least, one more extra spatial dimension. Along the large extra spatial dimensions there may be other universes. If some of them had the same laws of physics as ours, one can speculate about advanced civilizations able to travel through extra dimensions for visitation or colonization purposes, in either direction.

1 The Fermi Paradox

Los Alamos, summer 1950. At lunch Enrico Fermi, Edward Teller and other colleagues bring about the subject of unidentified flying objects, which was very popular at that time.

After a while, when they had changed subjects Fermi suddenly asked: Where is everybody? Performing fast mental computations, Fermi had reached the conclusion that alien civilizations should have been around visiting Earth for many thousands or millions of years. Therefore, why we do not see them? Although Fermi never explained how he made his computations, nor he gave an estimate of the number of civilizations which should have visited Earth, he had to rely on arguments like these: In our galaxy there are thousands of millions of stars much older than the Sun, many of them thousands of millions of years older. Therefore many civilizations must have arisen in our galaxy before ours and a fraction of them must have expanded through large regions of the galaxy or even through the whole galaxy. As a matter of fact, recently it has been found that the stars in the 'habitable zone' of the galaxy are in average one thousand million years older than the Sun [1].

Some other arguments that probably were not available at that time involve estimates about the lifetime of the second generation stars, inside of which the chemical elements of the organic matter are made, and also estimates about the total time necessary for a technological civilization to colonize, or visit, the whole galaxy. Regarding the second generation stars, they are formed only two million years after the supermassive first generation ones. The reason is that supermassive stars burn out completely exploding as supernovae in one million years only and it takes another million years for the debris to form new stars. As to the total time to colonize, or visit, the whole galaxy by a technological civilization, conservative computations of diffusion modeling give estimates from 5 to 50 million years [2], which is a cosmologically short timescale. Besides these considerations, the fact that life on Earth started very early supports the views, held by many scientists, that life should be abundant in the Universe. For example, in the opinion of the biochemist Christian de Duve, 'Life is almost bound to arise ... wherever physical conditions are similar to those that prevailed on our planet some four billion years ago'.

2   Solutions to the Fermi Paradox

Many solutions have been proposed to the Fermi paradox. We classify them as expansionist and non-expansionist, depending on whether they rely on the idea that technological civilizations expand or do not expand through large regions of the galaxy. The most popular non-expansionist solutions, based on the assumption that technological civilizations do not expand beyond a small neighborhood are the following ones:

- Interstellar travel is not possible no matter the scientific and technological level reached by a civilization. Advocates of this idea are, for example, most experts of the SETI (Search for Extraterrestrial Intelligence) project, who for about 30 years are trying to detect electromagnetic signals from distant civilizations.

- Generically, advanced civilizations have little or no interest in expanding through large regions of the galaxy.

- Technological civilizations annihilate themselves, or disappear by natural catastrophes, before having the chance to expand through large regions of the galaxy.

On the other hand, the most popular expansionist solutions to the Fermi Paradox, based on the assumption that generically technological civilizations (or a non-negligible fraction of them) do expand through large regions of the galaxy are the following ones:

- Alien civilizations do visit Earth at present times, for different purposes, and/or have visited Earth in the past. In this respect it is remarkable the fact that Francis Crick, one of the discoverers of the DNA structure, proposed in the mid-seventies that life on Earth could have been inseminated on purpose by alien intelligences. Besides, some scientists as well as many authors of popular books, have speculated that some unidentified flying objects could be true alien spacecrafts whereas some 'gods' descending from the sky, in many ancient traditions, could have been just alien astronauts (see for example [3]).

- Advanced alien civilizations have not encountered the Solar System yet, but they are on their way.

- Advanced alien civilizations might have strong ethical codes against interfering with primitive life-forms [4].

- Advanced aliens ignore us because of lack of interest due to our low primitive level. For example Robert Jastrow, ex-director of Mt. Wilson Observatory, claims [5] that, in average, advanced civilizations should consider us as larvae due to the fact that they should be thousands of millions of years ahead of us.... and who would be interested in communicating with larvae?

- Alien civilizations have not reached us yet because intelligent life is extremely difficult to emerge. As a result we could find ourselves among the most evolved technological civilizations in our galaxy or we could even be the only one. Gerard 't Hooft describes very sharply an extreme expansionist version of this view with the statement: 'Life on Earth must be a highly rare accident, in our galaxy at least, otherwise alien civilizations would be here, as suggested by the Fermi Paradox' [6].

Besides these simple solutions there are many more exotic proposals* For example, a rather drastic expansionist solution is given by the string theoretist Cumrun Vafa who thinks that the fact that we do not see aliens around could be the first proof of the existence of brane worlds: all advanced aliens would have emigrated to better parallel universes (our Universe has zero measure) [8].

Recently we made our own proposal for solving the Fermi Paradox [9]. It states that, at present, all the typical galaxies of the Universe are already colonized (or large regions of them) by advanced civilizations, a small proportion of their individuals belonging to primitive subcivilizations, like ours. That is, we put forward the possibility that our small terrestrial civilization is embedded in a large civilization unknowingly and this situation should be common in all typical galaxies. Whether the primitive subcivilizations would know or ignore their low status would depend, most likely, on the ethical standards of the advanced civilization in which they are immersed. If the standards were low, the individuals of the primitive subcivilizations would be surely abused in many ways. Consequently, in this case the primitive individuals would be painfully aware of their low status. If the ethical standards of the advanced individuals were high instead, then very probably they would respect the natural evolution (social, cultural) of the primitive subcivilizations, treating them 'ecologically' as some kind of protected species. In this case, which could well describe the situation of the terrestrial civilization, the primitive individuals would be completely unaware of the existence of the large advanced civilization in which they are immersed. Observe that the 'alien visitors', from the viewpoint of the primitive individuals, would not be so from the viewpoint of the advanced individuals because they rather would be visiting, or working in, their own territory. Observe also that we do not postulate that advanced alien civilizations might have strong ethical codes against interfering with primitive civilizations. We simply distinguish between aggressive and non-aggresive advanced civilizations, which in our opinion is a much more realistic idea. The fact that our civilization has never been attacked by aggressive aliens, as far as history knows, could be a clue that we belong to a non-aggressive advanced civilization which protects planet Earth, as part of its territory.

If this scenario were true for our civilization, then the *Subanthropic Principle* [9] would also hold. It states that we are not typical among the intelligent observers from the Universe. Typical civilizations of typical galaxies would be hundreds of thousands, or millions, of years more evolved than ours and, consequently, typical intelligent observers would be orders of magnitude more intelligent than us. In order for our proposal to be a solution of the Fermi Paradox, we complement it with an additional hypothesis, called the *Undetectability Conjecture* [9], which explains why we do not detect any signals of civilization from the

---

*As many as fifty solutions to the Fermi Paradox have been collected in the book [7], although there are several left out which we mention in this article.

outer space. This conjecture states that, generically, all advanced enough civilizations camouflage their planets for security reasons, because of the existence of aggressive advanced civilizations, so that no signal of civilization can be detected by external observers, who would only obtain distorted data for disuasion purposes.

3 The Fermi Paradox in the light of the inflationary and brane world cosmologies

3.1 Inflationary cosmologies

Almost two years ago Ken D. Olum argued [10] that in the infinite Universe predicted by eternal inflation there must be some large civilizations which have spread across their galaxies and contain a huge number of individuals. Although the Fermi Paradox was not mentioned, the underlying idea was again that in the observable Universe, because of the existence of thousands of billions of stars much older than the Sun, there must be large civilizations much older than ours. Then the author presented some computations regarding the probabilities that typical intelligent observers belong to a large (galactic size) civilization at the present time. In particular, using the assumption of an infinite Universe, like in the models of eternal inflation, and doing some conservative computations he predicted that '*all but one individual in 10^8 belongs to a large civilization*'.

Dropping the infinite Universe assumption, but keeping still inflation, the author claimed that the predictions are not very different than for the previous case because inflationary models, even if not eternal, usually produce a Universe much larger that the Universe we observe. Then, invoking the anthropic premise that we are typical individuals, he predicted that there is a probability of 10^8 versus 1 that we belong to a large civilization, in conflict with observation. The author concluded: '*A straightforward application of anthropic reasoning and reasonable assumptions about the capabilities of other civilizations predict that we should be part of a large civilization spanning our galaxy. Although the precise confidence to put in such a prediction depends on one's assumptions, it is clearly very high. Nevertheless, we do not belong to such a civilization. Thus something should be amiss....... but then what other mistakes are we making.....?*' According to our proposal we could be part of a large civilization spanning our galaxy, or a large region of it, without being aware, because of our primitive low status together with the high ethics of our hosts, as discussed in the previous section. The two major mistakes of the author, therefore, would have been to assume: first, that we are typical intelligent observers, and, second, that to belong to a civilization implies to be a citizen of it.

Olum's article seems to make the Fermi Paradox even stronger. However we do not agree with these views because, in our opinion, only our own galaxy matters for the problem at hand, regarding the observable Universe. The reason is that any other galaxies are much too far to be even considered as candidates for visitation or colonization in either direction (our closest neighbour Andromeda is two million light years away). As a result, it should not matter whether or not there is inflation or whether or not the Universe is infinite or finite, as long as inflation and the age of the Universe have little influence on the age of our own galaxy, in particular on the age of the Sun and the thousands of millions of stars older than the Sun. In other words, cosmological inflation affects the large scale structure of the Universe, but not the small galaxy-size scale, which is the only relevant scale for the spreading of civilizations. Otherwise one would have to postulate the existence of exotic phenomena, like wormholes [11], Alcubierre warp drive [12], or Krasnikov tubes [13], within the context of General Relativity, or some kind of tunneling, in order to connect efficiently spatially remote places, although the engineering of such solutions encounters seemingly insurmountable obstacles, such as unattainable energy requirements [14] and the need for exotic matter.

3.2 Brane world cosmologies

In the last years brane world models have been of increasing interest for both Particle Physics and Cosmology. They put forward the possibility that our Universe is located in a subspace (brane) of a higher dimensional spacetime, with the standard model fields confined on the brane and only gravity propagating in the bulk. This allows large, and even infinite, extra dimensions. Recent work on brane worlds started following the proposals of N. Arkani-Hamed, S. Dimopoulos and G.R. Dvali [15], and L. Randall and R. Sundrum [16]. The interest in these models comes from the fact that they offer the possibility to solve, or view from a newly different perspective, many longstanding problems in Particle Physics and Cosmology. First of all brane worlds may solve the hierarchy problem between the electroweak scale M_W and the Planck scale M_{Pl} without the need to introduce supersymmetry: M_W would be the only fundamental scale in nature, and the weakness of the 4-dimensional gravity would be just a consequence of the 4-dimensional graviton wave function being diluted in the bulk. But brane worlds can also shed some light on the baryogenesis and leptogenesis, on the proton stability, on the small masses and large mixing of the neutrinos, on the gravitational lensing (by brane world black holes), on the nature of the dark matter and dark energy, etc. (see [17] for brane world reviews).

In contrast with the inflationary cosmologies, brane world cosmologies have the potential to truly strengthen the Fermi Paradox. The reason is that in the brane world scenarios our observable Universe is embedded in a much larger cosmos with, at least, one more large extra spatial dimension. Along the large extra spatial dimensions there may be other universes which could be parallel to our own, or intersecting it somewhere. If any of these scenarios turns out to describe the real world, then it would be natural to expect that some of these universes would have the same laws of physics as ours and many of the corresponding advanced civilizations would master techniques to travel or 'jump' through the extra dimensions. This opens up enormous possibilities regarding the expansion of advanced civilizations simultaneously through several parallel universes with the same laws of physics, resulting in multidimensional empires. It could even happen that the expansion to other parallel galaxies through extra dimensions could be easier, with lower cost, than the expansion inside one's own galaxy.

In many other universes, however, the laws of physics would be different, corresponding perhaps to different vacua of the 'would be' ultimate Theory of Everything, resulting probably in 'shadow matter' universes with respect to ours. This means that shadow matter would only interact with our matter gravitationally, in the case it would be brought to our Universe using appropiate technology. This does not mean, however, that the shadow universes would be necessarily empty of intelligent beings. If some of them had advanced civilizations, some of their individuals could even 'jump' to our Universe, but not for colonization purposes since they would not even see our planets and stars, which they would pass through almost unaware (they would only notice the gravitational pull towards their centers). And the other way around, we could neither see, nor talk to, the shadow visitors, although they could perhaps try to communicate with the 'would be' intelligent beings of our Universe through gravitational waves, for example.

At present we are still in a very premature phase in the study of brane worlds and we do not know whether these ideas are in fact realistic. Nevertheless, the idea of large extra dimensions and parallel universes is acquiring bigger and bigger momentum in the scientific community, among both theoreticians and experimentalists. As a matter of fact, experimental signatures of large extra dimensions at present and future colliders are well understood by now [18] and an intense experimental search is currently under way. For example, experiments starting in 2007 at LHC (CERN) will be looking, among other things, for signatures of large extra dimensions.

4 Conclusions and Final Remarks

We have discussed whether the inflationary and brane world cosmologies have the potential to influence the problem known as the Fermi Paradox. We conclude that cosmological inflation does not influence the spreading of civilizations through our galaxy, contrary to recent claims, since it does not affect the age of the stars in it. Even if inflation produces a much older, or infinite Universe, like in eternal inflation models, it will be only relevant at very large scales far beyond visitation or colonization by technological civilizations, unless one invokes very exotic phenomena capable to connect distant galaxies efficiently enough with regards to space travel. Only in this case an older, or infinite, Universe would increase the probabilities of visitation by civilizations of other galaxies in our observable Universe.

In the case of brane world cosmologies we conclude that some of these scenarios could have the potential to strengthen the Fermi Paradox, provided they involve parallel universes with the same laws of physics as ours. Then it would be natural to expect the existence of advanced civilizations capable of traveling through extra dimensions for visitation or colonization purposes, in either direction. It could even happen that the expansion to other parallel galaxies through extra dimensions could be easier, with lower energetic cost, than the expansion inside one's own galaxy.

Acknowledgements

I am indebted to Pedro González-Díaz for several interesting discussions.

References

[1] C.H. Lineweaver, Y. Fenner, and B.K. Gibson, *'The galactic habitable zone and the age distribution of complex life in the Milky Way'*, Science 303, 59, 2004.

[2] I. Crawford, *'Where are they?'*, Scientific American, July 2000

[3] J.W. Deardorff, *'Possible extraterrestrial strategy for Earth'*, Q. J. R. Astron. Soc., 27, 94, 1986.

[4] W.I. Newman and C. Sagan, *'Galactic civilizations: Population dynamics and interstellar diffusion'*, Icarus 46, 293, 1981.

[5] R. Jastrow in videos of the series *'The Universe of Stephen Hawking'*, 1996.

[6] G. 't Hooft, private communication.

[7] S. Webb, *'If the Universe is Teeming with Aliens.... Where is Everybody? Fifty Solutions to the Fermi Paradox and the Problem of Extraterrestrial Life '*, Copernicus Books, New York, 2002.

[8] C. Vafa, private communication.

[9] B. Gato-Rivera, *'Brane Worlds, the Subanthropic Principle and the Undetectability Conjecture'*, physics/0308078, 2003.

[10] K. Olum, *'Conflict between anthropic reasoning and observation'*, *ANALYSIS* 64, 1, 2004.

[11] M.S. Morris and K.S. Thorne, *Am. J. Phys.* 56, 395, 1988.
M. Visser, *Lorentzian Wormholes: From Einstein to Hawking*, AIP Press, Woodbury, New York, 1996.
M. Visser, S. Kar and N. Dadhich, *Phys. Rev. Lett.* 90, 201102-1, 2003

[12] M. Alcubierre, *Class. Quant. Grav.* 11, L73, 1994.
H.E. Puthoff, *Phys. Essays* 9, 156, 1996.

[13] S.V. Krasnikov, *Phys. Rev. D* 57, 4760, 1998.

[14] M.J. Pfenning and L.H. Ford, *Class. Quant. Grav.* 14, 1743, 1997.

[15] N. Arkani-Hamed, S. Dimopoulos and G.R. Dvali, *Phys. Lett. B* 429, 263, 1998; *Phys. Rev. D* 59, 86004, 1999; *Phys. Today* 55N2, 35, 2002.

[16] L. Randall and R. Sundrum, *Phys. Rev. Lett.* 83, 4690, 1999.

[17] R. Maartens, *Living Rev. Rel.* 7, 1, 2004.
I. Antoniadis, *Eur. Phys. J. C* 33, S914, 2004.
C. Csaki, *'TASI lectures on extra dimensions and branes'*, hep-ph/0404096.
V. Rubakov, *Phys. Ups.* 44, 871, 2001.
D. Langlois, *Prog. Theor. Phys. Suppl.* 148, 181, 2003.

[18] G.F. Giudice, R. Rattazzi and J.D. Wells, *Nucl. Phys. B* 544, 3, 1999.
E. A. Mirabelli, M. Perelstein and M.E. Peskin, *Phys. Rev. Lett.* 82, 2236, 1999.
K. M. Cheung and W.Y. Keung, *Phys. Rev. D* 60, 112003, 1999.

In: Trends in General Relativity and Quantum Cosmology
Editor: Charles V. Benton, pp. 37-48

ISBN 1-59454-794-7
© 2006 Nova Science Publishers, Inc.

Chapter 3

ON $3+1$ DIMENSIONAL SCALAR FIELD COSMOLOGIES

F.L. Williams and P.G. Kevrekidis
Department of Mathematics and Statistics,
University of Massachusetts, Amherst MA 01003-4515, USA
T. Christodoulakis, C. Helias, and G.O. Papadopoulos
Nuclear and Particle Physics Section, Physics Department,
University of Athens, Panepistimiopolis, Ilisia, Athens 15771, Greece
Th. Grammenos
Dept. of Mechanical and Industrial Engineering,
University of Thessaly, Volos 38334, Greece

Abstract

In this communication, we analyze the case of $3+1$ dimensional scalar field cosmologies in the presence, as well as in the absence of spatial curvature, in isotropic, as well as in anisotropic settings. Our results extend those of Hawkins and Lidsey [Phys. Rev. D **66**, 023523 (2002)], by including the non-flat case. The Ermakov-Pinney methodology is developed in a general form, allowing through the converse results presented herein to use it as a tool for constructing new solutions to the original equations. As an example of this type a special blowup solution recently obtained in Christodoulakis *et al.* [gr-qc/0302120] is retrieved. Additional solutions of the 3+1 dimensional gravity coupled with the scalar field are also obtained. To illustrate the generality of the approach, we extend it to the anisotropic case of Bianchi types I and V and present some related open problems.

1 Introduction

In the past few years, the ekpyrotic scenario has been proposed as an alternative to the standard inflationary cosmology [1]. In this interpretation, the big bang is viewed as the collision of two domain walls or branes described by Einstein's equations coupled to a scalar field. The scalar in this case parametrizes the separation between the branes. Hence, in this dynamical setup, it is of interest to understand the coupling of scalar fields to gravity.

This role has been investigated rather extensively in 2+1 dimensional setups; see e.g., the earlier works of [2, 3, 4]. More recently the 3+1 dimensional case has become of interest; see e.g., [5, 6, 7, 8].

These studies have in fact motivated the present work. In particular, in a significant recent paper [5], Hawkins and Lidsey have proposed a connection between the so-called Ermakov-Pinney (EP) equation and flat, 3+1 dimensional, scalar field cosmologies. This is of particular interest since the EP equation is a very special, linearizable, nonlinear ordinary differential equation (ODE) that can be solved exactly if the underlying linear Schrödinger equation can be solved. We note in passing that one of the authors of [5] went on to use this approach to illustrate analogies between scalar field cosmological models and the dynamics of moments of the wavefunction of Bose-Einstein condensates [9]. This point further highlights the importance of this ODE that has been recurring in a variety of different areas as diverse as nonlinear optics [10], elasticity [11], quantum field theory [12] or molecular physics [13]. For a recent review of the EP equation and its applications, see e.g., [14].

One of the purposes of the present short communication is to demonstrate how the methodology developed in [5] can be generalized in the case of non-zero curvature. We will also explicitly provide a converse result according to which, given a solution of the EP equation, the solution of a corresponding Einstein equation, coupled to a scalar field, can be derived. The use of the converse result will be explicitly demonstrated in a special case, that of a linear Klein-Gordon equation for the scalar field. The special solution for this case provided recently in [6] will be retrieved. More general solutions will also be constructed.

As our gravitational model, we will use the 3-dimensional Friedmann-Robertson-Walker (FRW) metric. While other models such as the Brans-Dicke metric, see e.g., [15], are also popular, the FRW metric is widely accepted in the early universe scenario that is of primary interest herein (see the references mentioned above). The relevant line element (in co-moving coordinates and with a "cosmological" time choice) will thus read:

$$ds^2 = -dt^2 + a^2(t)\left(\frac{1}{1-cr^2}dr^2 + r^2 d\theta^2 + r^2 \sin^2\theta d\phi^2\right) \quad (1)$$

In the metric of Eq. (1), a is the scale factor while c describes the curvature of the spatial slice and can be normalized to the values $-1, 0, 1$ in the hyperbolic, flat and elliptic case respectively.

However, the above setting seems to restrict our considerations to an isotropic scenario. To illustrate the generality of the EP reduction and the usefulness of the corresponding technology, we also examine in the present work, an anisotropic case, namely the one of Bianchi types I and V. In the latter, spatially homogeneous, yet anisotropic geometries, the line element is given by:

$$ds^2 = -N(t)^2 dt^2 + \gamma_{\alpha\beta}(t)\sigma_i^\alpha(x)\sigma_j^\beta(x)dx^i dx^j \quad (2)$$

in the time gauge $N(t) = \sqrt{\gamma(t)}$; σ_i^α are the invariant basis one-forms of the homegeneous surfaces of simultaneity Σ_t, while $\gamma_{\alpha\beta}$ are the scale factors that constitute the (in principle) dynamical variables of the metric. We restrict ourselves in the present study to these anisotropic models, as these have been argued to have the desirable property of isotropizing at arbitrarily long times [16]. Additionally, type V is the simplest Bianchi model that admits velocities or tilted sources, hence it is natural to consider it as the simplest extension that would allow the universe to choose a reference frame at the exit from inflation (given that the de Sitter metric does not have a preferred frame).

Our presentation will proceed as follows: in section 2 we will provide the general EP methodology for the FRW system in the presence of a scalar. In section 3, we will apply these results in the special case of a massive scalar. In section 4 we will give additional special solutions to these equations, while in section 5, we will apply the method to the anisotropic case of Bianchi types I and V. Finally, in section 6, we will summarize our findings and present our conclusions.

2   3+1 Isotropic Scalar Field with Curvature: the EP reduction

We will follow the notation of [5] in what follows. In particular, in this setting, the Einstein equations of gravity and the Klein-Gordon equation for the scalar field can be respectively written as

$$H^2 + \frac{c}{a^2} = \frac{\kappa^2}{3}\left[\frac{1}{2}\dot{\phi}^2 + V(\phi) + \frac{D}{a^n}\right] \tag{3}$$

$$\ddot{\phi} + 3H\dot{\phi} + \frac{dV}{d\phi} = 0. \tag{4}$$

$H = \dot{a}/a$ represents the Hubble parameter. The first two terms in the bracketed expression of Eq. (3) denote the energy density of the scalar field with potential $V(\phi)$. The last term is the density of matter for the barotropic fluid with equation of state $p_{mat} = (n-3)\rho_{mat}/3$; $D \geq 0$, $0 \leq n \leq 6$, $\kappa^2 = 8\pi/m_P^2$, where m_P is the Planck mass. Finally, the dots will be used for differentiation with respect to the cosmological or coordinate time t. Notice that with respect to the corresponding equations (2)-(3) of [5], the former is augmented by the spatial curvature term (recall that c can take the values $-1, 0$ and 1 depending on the curvature of the hypersurface $t =$const.) in the left hand side.

Now, by using a further differentiation of Eq. (3) and the substitution $b = a^{n/2}$ (cf. with Eq. (8) of [5]), we obtain:

$$\frac{2}{n}\left(\frac{\ddot{b}}{b} - \frac{\dot{b}^2}{b^2}\right) - \frac{c}{b^{4/n}} = -\frac{\kappa^2}{2}\left[\dot{\phi}^2 + \frac{nD}{3b^2}\right]. \tag{5}$$

If we now define a new comoving time τ such that $\dot{\tau} = b$, then Eq. (5) becomes

$$\frac{d^2 b}{d\tau^2} + \frac{\kappa^2 n}{4}\left(\frac{d\phi}{d\tau}\right)^2 b = -\frac{\kappa^2 n^2 D}{12 b^3} + \frac{nc}{2 b^{\frac{4+n}{n}}}, \tag{6}$$

where the chain rule has been used and the functions in eq. (5) have been assumed as $b = b(t(\tau))$ and $\phi = \phi(t(\tau))$. This equation can be compared with Eqs. (10)-(11) of [5], with the difference being evident in the inclusion of the curvature term (the last term on the right hand side of Eq. (6)).

This prompts us to (briefly) discuss the EP equation which naturally arises in this context not only in the flat case of $c = 0$ examined in [5], but also in the case of $n = 2$. The latter is a remarkable example of a nonlinear yet integrable ordinary differential equation (ODE) of the form:

$$Y'' + Q(\tau)Y = \frac{\lambda}{Y^3}. \tag{7}$$

The particularly appealing feature of this nonlinear ODE is that its general solution can be obtained, provided that one is able to solve the linear Schrödinger (LS) problem $Y'' + Q(\tau)Y = 0$. For details on the properties of the EP equation, the interested reader is referred to [5, 14] and references therein. Here we just mention its basic superposition principle property. Namely, if the linearly independent solutions of the LS equation are $Y_1(\tau)$ and $Y_2(\tau)$, then the most general possible solution of the EP equation is given by

$$Y(\tau) = \left(AY_1^2 + BY_2^2 + 2CY_1Y_2\right)^{1/2} \tag{8}$$

where A, B and C are constants connected through

$$AB - C^2 = \frac{\lambda}{W^2} \tag{9}$$

and the Wronskian $W = Y_1 Y_2' - Y_2 Y_1'$.

It is of particular interest to note that the result of Eq. (6) can be used conversely for constructing solutions of scalar field cosmologies, using the EP equation structure and solutions. The converse result can be proved in the following form: Given Q and $\lambda = -\kappa^2 n^2 D/12 < 0$, let $Y > 0$ be a solution of

$$\frac{d^2Y}{d\tau^2} + QY = \frac{\lambda}{Y^3} + \frac{nc}{2Y^{1+\frac{4}{n}}} \tag{10}$$

Define a new time coordinate t such that $\dot{\tau} = Y(\tau(t))$ and $a = Y^{2/n}$, as well as a new field ϕ satisfying:

$$\frac{n\kappa^2}{4}\left(\frac{d\phi}{d\tau}\right)^2 = Q, \tag{11}$$

with $Q \neq 0$. Finally, define a potential:

$$V(\phi) = \frac{12}{\kappa^2 n^2}\left(\frac{dY}{d\tau}\right)^2 - \frac{Y^2}{2}\left(\frac{d\phi}{d\tau}\right)^2 - \frac{D}{Y^2} + \frac{3c}{\kappa^2 Y^{4/n}}. \tag{12}$$

Then, if we consider the triplet $(a(\tau(t)), \phi(\tau(t)), V(\phi))$, the latter satisfies the Eqs. (3)-(4).

Let us now give a number of specific examples, where this construction scheme can be used to obtain solutions to the $3+1$ dimensional scalar field cosmology equations.

3 Applications

3.1 Flat FRW Metric and Massless Scalar

In the absence of matter, we can set $D = 0 \Rightarrow \lambda = 0$, and consider for convenience the case of $n = 2$. For $\lambda = 0$, and for the flat FRW metric e.g., for $c = 0$, Eq. (10) becomes

$$\frac{d^2Y}{d\tau^2} + QY = 0 \tag{13}$$

Assuming the general case $V(\phi) = m^2\phi^2/2$, the particular scenario of a massless scalar yields $V(\phi) = 0$, hence Eq. (12) becomes:

$$\frac{d^2Y}{d\tau^2} + \frac{3}{Y}\left(\frac{dY}{d\tau}\right)^2 = 0 \qquad (14)$$

whose solution yields:

$$Y(\tau) = A\tau^{1/4}; \qquad (15)$$

hence from Eq. (13):

$$Q(\tau) = \frac{3}{16}\frac{1}{\tau^2}. \qquad (16)$$

This, in turn, through the definition of $\dot{\tau}$ yields:

$$\tau(t) = \left(\frac{3}{4}\right)^{\frac{4}{3}} A^{4/3} t^{\frac{4}{3}} \qquad (17)$$

and, thus, finally from Eq. (11):

$$\phi(t) = \sqrt{\frac{2}{3}}\log(\frac{3A}{4}) + \sqrt{\frac{2}{3}}\log(t - t_0). \qquad (18)$$

which is the same solution as obtained in Eq. (8) of [6].

3.2 A Generalization: Non Flat FRW Metric Coupled with Scalar

To demonstrate the generality of the technique also in non-flat cases with $c \neq 0$, we examine the example of $n = 2$, $\kappa = 1$ and $D > c/3$ (which implies that $\lambda + c < 0$).

In this case, Eq. (10) becomes:

$$\frac{d^2Y}{d\tau^2} + QY = \frac{\lambda + c}{Y^3} \qquad (19)$$

Motivated by cases in which we are able to solve the EP (or equivalently the underlying linear Schrödinger) equation exactly, we choose $Q(\tau) = 3/(16\tau^2)$ (cf. Eq. (16)). In this case, using Eq. (8), we can find the general solution to Eq. (19) as:

$$Y(\tau) = \left(A\tau^{\frac{3}{4}} + B\tau^{\frac{1}{4}} + 2C\tau\right)^{\frac{1}{2}}, \qquad (20)$$

where $AB - C^2 = 4(\lambda + c)$. For convenience, we use our freedom of coefficients to set $A = B = 0$, hence $C = 2\sqrt{|\lambda + c|}$. We thus obtain:

$$Y(\tau) = (2C\tau)^{\frac{1}{2}}; \qquad (21)$$

From Eq. (11), we derive:

$$\phi(\tau) = \frac{\sqrt{6}}{4}\log(\tau). \qquad (22)$$

From the definition of $\dot{\tau}$, it can be seen that:

$$\tau = \frac{C}{2}(t-t_0)^2. \tag{23}$$

Finally, from Eq. (12), it follows that:

$$V(\phi) = \frac{S}{\tau} \equiv Se^{-\frac{4}{\sqrt{6}}\phi} \tag{24}$$

where $S = 9c/8 - D/(2c) + 3/2$.

This result generalizes the corresponding result of Eq. (31) of [5] in the case in which curvature is present.

3.3 An Example of Quadratic Scalar Fields

As another example, we will use a case in which we use the EP approach in a reverse engineering way. In particular, in Eq. (10), we will postulate the solution and we will obtain the potential that is compatible with this solution. We assume that matter is absent, hence $D = \lambda = 0$, for $n = 4$ and $c \geq 0$. We now demand that $Y(\tau) = 2B\tau$. Then

$$Q(\tau) = \frac{c}{4B^3\tau^3}. \tag{25}$$

From the definition of $\dot{\tau}$, we obtain that

$$\tau(t) = \frac{A^2}{2B}e^{2Bt}, \tag{26}$$

where A is (without loss of generality) a positive constant. Then from Eq. (11), we obtain:

$$\phi(t) = -\frac{\sqrt{2c}}{\kappa AB}e^{-Bt} + \alpha, \tag{27}$$

where α is an arbitrary constant, while the potential

$$V(\phi) = \frac{3B^2}{\kappa^2} + B^2(\phi - \alpha)^2 \tag{28}$$

is quadratic in ϕ. Notice that in this example, the restriction of $n = 4$ can be lifted: in particular substituting $A \to A^{n/4}$ and $B \to Bn/4$, we can perform the same calculation for any power n, but the expression for $V(\phi)$ is functionally the same as the one of Eq. (28).

We remark here that the solution found in this subsection was previously identified by means of a different (than the EP) approach in [17] (cf. Eqs. (24)-(28) therein).

3.4 An Example of Constant Scalar Fields

In the same inverse procedure spirit, another convenient choice of $Q(\tau)$ is $Q(\tau) = 0$ (for $n = 2$). Then the EP equation has the straightforward general solution:

$$Y(\tau) = \left(A\tau^2 + B + 2C\tau\right)^{1/2}, \tag{29}$$

with

$$AB - C^2 = \lambda + c = -\frac{\kappa^2 D}{3} + c \equiv \tilde{\lambda}. \tag{30}$$

Consequently,

$$\tau(t) = \frac{D}{4A} \exp\left(A^{1/2}t\right) + \frac{\tilde{\lambda}}{DA^{1/2}} \exp\left(-A^{1/2}t\right) - \frac{C}{A}. \tag{31}$$

It can then be immediately seen that the scalar ϕ and the potential V are constants, while the scale factor

$$a(t) = a(0) \cosh\left(A^{1/2}t\right) + \sqrt{a(0)^2 - \frac{\tilde{\lambda}}{A}} \sinh\left(A^{1/2}t\right) \tag{32}$$

with

$$a(0) = \frac{D}{4A^{1/2}} + \frac{\tilde{\lambda}}{DA^{1/2}}. \tag{33}$$

It should be noted that this is a generalization of the solution of Eq. (24) of [5] in the spatially non-flat case. Similar solutions have been obtained through direct calculations (i.e., instead of the EP methodology) in [3].

3.5 Obtaining 3+1 dimensional solutions from 2+1 dimensional solutions

Let us now assume that (a_2, ϕ_2, V_2) is a solution of the 2+1-dimensional version of the Einstein equation together with a Klein-Gordon for scalar field, i.e., they satisfy [4]

$$H_2^2 + \frac{c}{a_2^2} = G\left[\frac{1}{2}\dot{\phi}_2^2 + V_2(\phi_2) + \rho_m\right] \tag{34}$$

$$\ddot{\phi}_2 + 2H_2\dot{\phi}_2 + \frac{dV_2}{d\phi_2} = 0. \tag{35}$$

In these equations $c \in \{-1, 0, 1\}$, $H_2 = \dot{a}_2/a_2$ and, in general, the matter density $\rho_m = \rho_m(0)\left[a_2(0)/a_2(t)\right]^{2\gamma}$ (with $1 \leq \gamma \leq 2$). Then, it can be straightforwardly proved by matching of the corresponding terms that, given such a solution of the 2+1-dimensional system, a solution (a, ϕ, V) of Eqs. (3)-(4) of the 3+1-dimensional system can be constructed; see e.g., [2, 3, 4] and the specific example mentioned below for an illustration. More specifically, this is done using $a = a_2$, $\phi = \frac{\sqrt{2G}}{\kappa}\phi_2 + s$ (s is an arbitrary constant) and

$$V(\phi) = \frac{3G}{\kappa^2}\left[\frac{\left(\dot{\phi}_2(\psi(\frac{\kappa}{\sqrt{2G}}(\phi - s)))\right)^2}{6} + V_2\left(\frac{\kappa}{\sqrt{2G}}(\phi - s)\right)\right]; \tag{36}$$

ψ is the inverse of ϕ_2 i.e., $\psi(\phi_2(t)) = t$. Additionally, $n = 2\gamma$ and $\kappa^2 n^2 D = G\gamma^2 \rho_m(0) a_2(0)^{2\gamma}$, to make the matching between the solutions of Eqs. (34)-(35) and Eqs. (3)-(4) complete.

As an example, we consider the case of $c = 0$ and $\rho_m = 0$ (hence κ is arbitrary and $D = 0$) and use the solution of [2]

$$a_2 = t^2 \left[1 + \frac{A}{t^3}\right]^{1/2} \tag{37}$$

$$\phi_2 = \frac{1}{\sqrt{2G}} \log\left[\frac{Gt^2 \Lambda}{3}(1 + \frac{A}{t^3})\right] \tag{38}$$

$$V_2(\phi_2) = \Lambda e^{-\sqrt{2G}\phi_2}. \tag{39}$$

Then, the resulting solution of the 3+1-dimensional problem satisfies $a = a_2$, $\phi = \frac{\sqrt{2G}}{\kappa}\phi_2 + s$ and for the potential

$$V(\phi) = \frac{G^2 \Lambda^2}{36\kappa^2} \left[\frac{A^2 + 32A\chi^3 + 40\chi^6}{\chi^4}\right] e^{-2\kappa(\phi - s)}, \tag{40}$$

where

$$\chi = \left[-\frac{A}{2} + \sqrt{\frac{A^2}{4} - \frac{e^{3\kappa(\phi-s)}}{G^3 \Lambda^3}}\right]^{1/3} - \left[\frac{A}{2} + \sqrt{\frac{A^2}{4} - \frac{e^{3\kappa(\phi-s)}}{G^3 \Lambda^3}}\right]^{1/3}, \tag{41}$$

a result that is more explicit than the corresponding one of [7] in that it contains the solution of the relevant inverse problem for the function ψ.

4 Anisotropic Generalizations: EP reduction for Bianchi Types I and V

4.1 Bianchi Type I

Since the anisotropic case was not previously discussed in the earlier work of [3, 4, 5], we discuss it here in more detail.

In this case, the dynamical scale factors and the one-forms in the metric of Eq. (2) are given by:

$$\gamma_{\alpha\beta}(t) = \begin{pmatrix} A(t)^2 & 0 & 0 \\ 0 & B(t)^2 & 0 \\ 0 & 0 & \Gamma(t)^2 \end{pmatrix}, \quad \sigma_i^\alpha(x) = \begin{pmatrix} 1 & 0 & 0 \\ 0 & 1 & 0 \\ 0 & 0 & 1 \end{pmatrix}$$

We can define the tensor: $F_{\mu\nu} = G_{\mu\nu} - 8\pi T_{\mu\nu}$ where $G_{\mu\nu} = R_{\mu\nu} - \frac{1}{2}g_{\mu\nu}R$ is the Einstein tensor and $T_{\mu\nu} = \phi_{;\mu}\phi_{;\nu} - \frac{1}{2}g_{\mu\nu}(\phi^{;\alpha}\phi_{;\alpha} + m^2\phi^2)$ is the energy momentum tensor. Then, the quadratic constraint is the equation $F_0^0 = 0$, the kinematic equation is given by $F_1^1 = 0$, while the Klein-Gordon equation for the field is given by $\phi_{;\mu}^{;\mu} - m^2\phi = 0 \propto T_{;\nu}^{\mu\nu} = 0$. Notice additionally, that the two integrals of the motion, namely $I_1 = F_1^1 - F_2^2 = 0$ and $I_2 = F_1^1 - F_3^3 = 0$ yield $B(t) = A(t)e^{\kappa t/2}$ and $\Gamma(t) = A(t)e^{\lambda t/2}$.

Solving the Klein-Gordon equation for $\phi''(t)$ and substituting the result into $\partial_t F_0^0 = 0$ (as well as solving $F_0^0 = 0$ for ϕ and using the resulting expression in $\partial_t F_0^0 = 0$), one is led to a dynamical equation for the remaining scale factor $A(t)$ in the form:

$$\frac{\kappa\lambda}{4} + \frac{\kappa A'(t)}{A(t)} + \frac{\lambda A'(t)}{A(t)} + \frac{4 A'(t)^2}{A(t)^2} - \frac{\phi'(t)^2}{2} - \frac{A''(t)}{A(t)} = 0 \qquad (42)$$

Using now: $A(t) = Y(t)^{2/n}$ and a change of variable $\tau = \int^t \Omega(t')dt'$, we obtain:

$$\ddot{Y}(\tau) + \dot{Y}(\tau)\frac{\Omega'(t)}{\Omega(t)^2} - \frac{\dot{Y}(\tau)^2}{Y(\tau)}\frac{(6+n)}{n} - (\kappa+\lambda)\frac{\dot{Y}(\tau)}{\Omega(t)} - \frac{n\kappa\lambda}{8}\frac{Y(\tau)}{\Omega(t)^2} + \frac{nY(\tau)\dot{\phi}(\tau)^2}{4} = 0 \qquad (43)$$

Hence, a choice of time reparametrization according to:

$$\frac{\Omega'(t)}{\Omega(t)} = \kappa + \lambda + \frac{(6+n)}{n}\frac{Y'(t)}{Y(t)} \qquad (44)$$

(which leads to $\Omega(t) = \theta e^{(\kappa+\lambda)t}Y(t)^{(6+n)/n}$, where $\theta > 0$ is a constant of integration), results in the form:

$$\ddot{Y}(\tau) + QY(\tau) = \frac{\Gamma}{Y(\tau)^{1+12/n}}, \quad Q = n\frac{\dot{\phi}(\tau)^2}{4}, \quad \Gamma = \frac{n\kappa\lambda e^{-2(\kappa+\lambda)t}}{8\theta^2}. \qquad (45)$$

Eq. (45) is of the form of Eq. (6) and becomes an EP equation for the choice of $n = 6$, for $\kappa = -\lambda$.

In the more general case, the problem becomes extremely complex as t depends on τ through $d\tau/dt = \Omega(t)$ and Ω is itself a function of the solution. Typically, this problem will not be analytically tractable (if combined, it leads to an integro-differential equation for $Y(\tau)$). However, this can be used as an *inverse* problem: we can *postulate* $\tau(t)$, derive from it $\Omega(t)$ and $Y(t)$ and use them in Eq. (45) to derive the form of $\phi(t)$.

A simple example of the above methodology can be given as follows: let us assume that $\tau = e^{(\kappa+\lambda)t}/(\kappa + \lambda)$, then $\Omega(t) = e^{(\kappa+\lambda)t}$. Hence, choosing $\theta = 1$, for $n = 6$ (the EP-like case for Eq. (45)) $Y(t) = Y(\tau) = 1$. Then, using the Eq. (45)), we find that

$$\phi(\tau) = C + \sqrt{\frac{\kappa\lambda}{2(\kappa+\lambda)^2}}\log(\tau) \qquad (46)$$

and hence (e.g., choosing the integration constant $C = \log(\kappa+\lambda)\sqrt{\frac{\kappa\lambda}{2(\kappa+\lambda)^2}}$), $\phi(t) = \sqrt{\frac{\kappa\lambda}{2}}t$.

4.2 Bianchi Type V

In this case, the scale factors and one-forms are, in turn, given by:

$$\gamma_{\alpha\beta}(t) = \begin{pmatrix} B(t)^2 & 0 & 0 \\ 0 & \Gamma(t)^2 & 0 \\ 0 & 0 & A(t)^2 \end{pmatrix}, \quad \sigma_i^\alpha(x) = \begin{pmatrix} 0 & e^{-x} & 0 \\ 0 & 0 & e^{-x} \\ 1 & 0 & 0 \end{pmatrix}.$$

In type V, there exist three integrals of the motion, namely: $I_1 = F_1^1 - F_2^2 = 0$, $I_2 = F_1^1 - F_3^3 = 0$ and $I_3 = F_0^1 = 0$, which yield: $B(t) = A(t)e^{\kappa t/2}$ as well as: $\Gamma(t) = A(t)e^{-\kappa t/2}$.

Once again following the same path as before and substituting into $\partial_t F_0^0 = 0$, we obtain a single ODE for $A(t)$ of the form:

$$-6\kappa^2 A(t) A'(t) - 24 A(t)^5 A'(t) + \frac{96 A'(t)^3}{A(t)} - 12 A(t) A'(t) \Phi'(t)^2 - 24 A(t) A''(t) = 0 \quad (47)$$

Through a similar motivation as in the previous subsection, we use the change of variables: $A(t) = Y(t)^{2/n}$, alongside the time reparametrization $\tau = \int^t \Omega(t')dt'$, with $\Omega(t) = \theta Y^{(6+n)/n}$. This results in the final form:

$$\ddot{Y}(\tau) + QY(\tau) = -\frac{n\kappa^2}{8\theta^2} \frac{1}{Y(\tau)^{1+12/n}} - \frac{n}{2\theta^2} \frac{1}{Y(\tau)^{1+4/n}}, \quad (48)$$

where $Q = n\frac{\dot{\phi}(\tau)^2}{4}$, which is again a generalized form of an EP equation. This statement is made in the sense that while there is no choice of the exponent that yields a mere inverse cubic dependence on $Y(t)$ in the right hand side of Eq. (48), the typical scenario involves inverse power dependence in a manner similar to the EP equation.

While the resulting Eq. (48) is not directly of the form of the EP equation, its dynamics can be controllably tuned to be close to those of the EP solutions. In particular, it is clear that the asymptotic behavior of the equation will be similar to that of $\ddot{Y}(\tau) + QY(\tau) = -\frac{n\kappa^2}{8\theta^2} \frac{1}{Y(\tau)^{1+12/n}}$ for small initial data, while it will be close to that of $\ddot{Y}(\tau) + QY(\tau) = -\frac{n}{2\theta^2} \frac{1}{Y(\tau)^{1+4/n}}$ for the case of large initial data. Hence, the choice of $n = 6$ for small initial data and the one of $n = 2$ for large initial data yields EP behavior (this point has also been verified in numerical investigations of the equation not shown here). In the intermediate regime where there is competition between the two terms the EP results will not be immediately applicable and one should resort to numerical simulations of Eq. (48).

5 Conclusions

In this short communication, we have generalized the earlier work of [5] and of [4] in the direction of obtaining solutions to $3+1$ dimensional cosmological models for the Friedmann-Robertson-Walker case in a systematic way. The method re-casts the relevant ordinary differential equations into one of the Ermakov-Pinney type which is explicitly solvable. From the solutions of the resulting EP equation one can re-construct the solutions to the original cosmological model in a step-by-step inverse process that has been detailed. The present study generalizes that of [5] in that cases of non-zero curvature of the FRW metric can be also considered. It also extends the results of [4] in the $3+1$ dimensional context. The method can be used to derive a variety of solutions in the latter context, including ones considered previously in [2, 3, 6].

On the other hand, we have also illustrated the generality of our method, by exploring the potential of such Ermakov-Pinney reductions to systems with anisotropy. In particular, we have demonstrated that the case examples of Bianchi types I and V can be reduced to Ermakov-Pinney-like equations. These Bianchi types were chosen as prototypical examples where a velocity/tilt can be included and as examples of Bianchi types that may eventually isotropize. The EP reduction of these cases highlights the generality and usefulness of the procedure. However, by the same token and since the EP is the only one of these inverse power, nonlinear ordinary differential equations for which an explicit solution exists, it underlines the importance of understanding the behavior of such classes of equations. In particular, examining numerically their temporal evolution for a number of physically motivated cases would be a natural next step. Such studies are currently in progress and will be reported in a future work.

G.O. Papadopoulos is currently a scholar of the Greek State Scholarships Foundation (I.K.Y.) and acknowledges the relevant financial support. T Christodoulakis and G.O. Papadopoulos, acknowledge support by the University of Athens, Special Account for the Research Grant-No. 70/4/5000 This work was also partially supported by the Eppley Foundation for Research, NSF-CAREER and NSF-DMS-0204585 (PGK). We are thankful to J.E. Lidsey (private communication with T.C.) for useful discussions and for pointing out to us the connection between the results of Section 3.3 and those of [17].

References

[1] J. Khoury, B.A. Ovrut, P.J. Steinhardt and N. Turok, *Phys. Rev. D* **64**, 123522 (2001); J. Khoury, B.A. Ovrut, P.J. Steinhardt and N. Turok, hep-th/0109050.

[2] J.D. Barrow, A.B. Burd and D. Lancaster, *Class. Quantum. Grav.* **3**, 551 (1986).

[3] N. Cruz and C. Martínez, *Class. Quantum Grav.* **17**, 2867 (2000).

[4] F.L. Williams and P.G. Kevrekidis, *Class. Quantum Grav.* **20**, L177 (2003).

[5] R.M. Hawkins and J.E. Lidsey *Phys. Rev. D* **66**, 023523 (2002).

[6] T. Christodoulakis, C. Helias, P.G. Kevrekidis, I.G. Kevrekidis and G.O. Papadopoulos, in *Non Linear Waves: Classical and Quantum Aspects*, F. Kh. Abdullaev and V. V. Konotop (Eds.), Kluwer Academic Publishers (2004), pp 135-143; see also gr-qc/0302120.

[7] A.A. Garcia, M. Cataldo and S. del Campo, *Phys. Rev. D* **68**, 124022 (2003).

[8] P.S. Apostolopoulos and M. Tsamparlis, *General Relativity and Gravitation* **36**, 277 (2004).

[9] J.L. Lidsey, *Class. Quantum Grav.* **21**, 777 (2004).

[10] A.M. Goncharenko, Yu.A. Logvin, A.M. Samson and P.S. Shapovalov, *Opt. Commun.* **81**, 225 (1991).

[11] M. Shahinpoor and J.L. Nowinski, *Int. J. Non-Linear Mech.* **6**, 193 (1971).

[12] F. Finelli, G.P. Vacca and G. Venturi, *Phys. Rev. D* **58**, 103514 (1998).

[13] Yu.B. Gaididei, K.Ø. Rasmussen and P.L. Christiansen, *Phys. Rev. E* **52**, 2951 (1995).

[14] P.B. Espinoza Padilla, math-ph/0002005.

[15] R. Carretero-González, H.N. Núñez-Yépez and A.L. Salas-Brito, *Phys. Lett. A* **188**, 48 (1994).

[16] C.B. Collins and S.W. Hawking, *Astrophys. J.* bf 180, 317 (1973).

[17] G.F.R. Ellis and M.S. Madsen, *Class. Quant. Grav.* **8**, 667 (1991).

Chapter 4

NOTES ON DILATON QUANTUM COSMOLOGY

Gabriel Catren
Instituto de Astronomía y Física del Espacio.
C.C. 67, Sucursal 28, 1428 Buenos Aires, Argentina.
E-mail:catren@iafe.uba.ar
Claudio Simeone
Departamento de Física, Facultad de Ciencias Exactas y Naturales,
Universidad de Buenos Aires.
Pabellón I, Ciudad Universitaria, 1428,
Buenos Aires, Argentina.
E-mail: csimeone@df.uba.ar

Abstract

In these notes we shall address the canonical quantization of the cosmological models which appear as solutions of the low energy effective action of closed bosonic string theory (dilaton models). The analysis will be restricted to the quantization of the minisuperspace models given by homogeneous and isotropic cosmological solutions. We shall study the different conceptual and technical problems arising in the Hamiltonian formulation of these models as a consequence of the presence of so called Hamiltonian constraint. In particular we shall address the problem of time in quantum cosmology, the characterization of the symmetry under clock reversals arising from the existence of a Hamiltonian constraint and the problem of imposing boundary conditions on the space of solution of the Wheeler–DeWitt equation.

1 Introduction

String cosmology received considerable attention in the last decade because of the new scenario that it proposes for the early universe [51, 52, 23]. When the high energy modes of the strings become negligible, the dynamical evolution of the universe begins to be dominated by the massless fields which act as the matter source of gravitational dynamics. This phase of the universe is commonly called the *dilatonic era*. The purpose of the present notes is to provide a consistent quantum description for this epoch. We shall address the formal aspects of the problem, more precisely, the obtention of a wave function allowing for a clear definition of probability within the context of the minisuperspace approximation.

In the minisuperspace picture all except a few degrees of freedom of the gravitational and matter fields are frozen at the classical level, so that the problem to be solved reduces from quantum field theory to quantum mechanics. Most developments in quantum cosmology have been achieved within the minisuperspace approximation. However, though the reduction to a finite number of degrees of freedom considerably simplifies the problem of obtaining a consistent quantum cosmology, the fact that the dynamical classical theory includes the general covariance as a central feature is an obstruction to a straightforward application of standard quantum mechanics.

This feature is most apparent when the classical theory of the gravitational field (even with the inclusion of matter) is formulated in the Hamiltonian form, as one immediately obtains that the dynamical evolution is governed by a Hamiltonian $\mathcal{H}$ which vanishes on the trajectories of the system [29, 6, 45]:

$$\mathcal{H} = G^{ik} p_i p_k + V(q) = 0.$$

Thus the theory involves a constraint which is quadratic in the momenta, which reflects the reparametrization invariance of the corresponding action (see below), i.e., the fact that the separation between successive spatial hypesurfaces in spacetime is arbitrary. In general, one also obtains linear constraints, analogous to those of Yang–Mills theories [6]. These linear constraints assure the invariance of the theory under a change of the spatial coordinates used to represent the spatial geometry of each hypersurface. We shall not discuss this point here, as their role is not central in the minisuperspace picture.

The quantization of constrained systems can be analyzed within both the canonical approach and the path integral formulation (see, for example, Ref. [31]). In the first of these frameworks, the direct application to cosmological models of the well known Dirac program [20] leads to the Wheeler–DeWitt equation [19], which is a second order equation in all the derivatives of the wave function Ψ. The solutions to this equation do not depend explicitly on the "time" parameter τ, but only on the coordinates, which reflects the reparametrization invariance of the classical theory. This reparametrization invariance means that the integration parameter τ is not a "true" time. This absence of a notion of time in the Wheeler–DeWitt quantization program is a serious obstacle for understanding the results in terms of conserved positive-definite probabilities [28, 7, 22]. Within dilaton cosmology this problem has received considerable attention (see, for example, Refs. [14, 13, 15, 16, 26]).

The canonical program admits an alternative approach to the Dirac method, which relies on the idea that it is necessary to "reduce" the theory before quantization, i.e., to find a "true" time at the classical level. If this reduction can be effectively performed, the original action can be reduced to an ordinary action without the reparametrization symmetry of the former. In that case the quantization program continues as if we were dealing with an ordinary classical theory. The theory can be then quantized by means of a Schrödinger equation with its typical conserved positive-definite probabilities.

Our aim will be then the analysis of these points by studying the cosmological models which appear as solutions of the low energy effective action of closed bosonic string theory [24, 23]. In Section 2 we shall begin by reviewing dilaton gravity, in particular homogeneous and isotropic cosmological solutions, whose formulation we shall put in Hamiltonian form. Then in Section 3 we shall address the problem of time and quantization, both in the usual Wheeler–DeWitt scheme as well as in the Schrödinger formulation, and we shall carefully discuss the relation between the corresponding solutions, their possible equivalence, and their role in selecting solutions with physical meaning (in particular when an extrinsic time must be introduced). Section 4 will be devoted to a more conceptual than technical discussion, with a wider scope which goes beyond the particular problem of dilaton models. Finally, in Section 5 we briefly summarize the essential points of the whole discussion.

2 Bosonic string theory and cosmological models

2.1 General theory

The action describing the world-sheet dynamics of strings on a curved manifold in presence of background fields has the form [39, 38]

$$S_{WS} = \frac{1}{4\pi\alpha'} \int d\sigma d\tau \sqrt{h} \left(h_{\alpha\beta} g_{\mu\nu}(X) + i\varepsilon_{\alpha\beta} B_{\mu\nu}(X) \right) \partial^\alpha X^\mu \partial^\beta X^\nu$$
$$+ \frac{1}{2\pi} \int d\sigma d\tau \sqrt{h} R(X) \phi(X), \tag{1}$$

where $h_{\alpha\beta}$ is the metric on the string world-sheet, R is the Ricci scalar related with this metric, $g_{\mu\nu}$ is the metric of the spacetime on which the theory is formulated, $B_{\mu\nu}$ is an antisymmetric field (commonly known as the NS-NS two-form field) and ϕ is the scalar dilaton field. These three fields appear in the massless spectrum of closed bosonic string theory. We have noted α, β for for the indices corresponding to the coordinates on the two-dimensional world-sheet, while the indices μ, ν, ρ correspond to the D-dimensional spacetime coordinates. The parameter α' (commonly known as the Regge slope) is the inverse of the string tension, $T = 1/(2\pi\alpha')$, which defines the scale of the theory at the quantum level. Clearly, the action (1) defines a two-dimensional field theory; this theory is invariant under the transformations

$$\delta B_{\mu\nu} = \partial_\mu \Lambda_\nu - \partial_\nu \Lambda_\mu, \qquad \delta\phi = \phi_0, \tag{2}$$

with Λ_μ an arbitrary vector and ϕ_0 a constant.

The action (1) is invariant under the Weyl –conformal– transformation $h_{\alpha\beta} \to \Omega^2(\tau, \sigma) h_{\alpha\beta}$. If we require that this symmetry holds also at the quantum level (no conformal anomalies) the beta functions must vanish [38, 23]; then at first order in the α' power expansion and introducing the strength tensor $H_{\mu\nu\rho}$ associated to the antisymmetric field $B_{\mu\nu}$

$$\mathbf{H}_{\mu\nu\rho} = \partial_\mu B_{\nu\rho} + \partial_\rho B_{\mu\nu} + \partial_\nu B_{\rho\mu}, \tag{3}$$

we obtain:

$$R_{\mu\nu} + \nabla_\mu \nabla_\nu \phi - \frac{1}{4} \mathbf{H}_{\mu\rho\delta} \mathbf{H}_\nu^{\rho\delta} = 0,$$
$$\nabla^\delta \mathbf{H}_{\delta\mu\nu} - \nabla^\delta \phi \mathbf{H}_{\delta\mu\nu} = 0,$$
$$c - \nabla_\mu \nabla^\mu \phi + \nabla_\mu \phi \nabla^\mu \phi - \frac{1}{6} \mathbf{H}_{\mu\nu\rho} \mathbf{H}^{\mu\nu\rho} = 0, \tag{4}$$

where, in principle, $c = 2(D-26)/(3\alpha')$. However, c can be changed by including more fields, so that in what follows we shall consider it as an arbitrary real number. Note then the difference between the theory for a point particle and the theory for a string: the quantum theory for the last one can not be consistently formulated in an arbitrary background, but, instead, it imposes restrictions on the admissible external fields.

Now, we are interested in a spacetime formulation of the theory of gravitation, analogous to the Einstein–Hilbert action that we have in General Relativity. It can be shown that the equations (4) can be interpreted as the Euler–Lagrange equations of motion of a field theory corresponding to the following effective action:

$$S_{SF} = \frac{1}{16\pi G_N} \int d^D x \sqrt{-g} e^{-\phi} \left(-c + R + \nabla_\mu \phi \nabla^\mu \phi - \frac{1}{12} H_{\mu\nu\rho} H^{\mu\nu\rho} \right), \quad (5)$$

where G_N is the D-dimensional Newton constant and R is the Ricci scalar of the spacetime. Thus, we can understand this as the low energy effective action describing the large tension limit ($\alpha' \to 0$) of closed bosonic string theory. A consistent configuration of background fields for the formulation of string theory must be a classical solution obtained from the variational principle corresponding to the action (5). In brief, the requirement of preserving at the quantum level the conformal invariance of the two-dimensional world sheet theory, formulated up to the first order in the inverse of the string tension, leads to the same equations of motion resulting from the variational principle $\delta S = 0$ imposed on the D-dimensional field theory given by (5).

A more familiar formulation can be obtained by redefining the fields as $g_{\mu\nu} \to e^\phi g_{\mu\nu}$, so that the action of the spacetime theory becomes

$$\begin{aligned} S_{EF} &= \frac{1}{16\pi G_N} \int d^D x \sqrt{-g} \\ &\times \left(R - c e^{2\phi/(D-2)} + \frac{1}{D-2} \nabla_\mu \phi \nabla^\mu \phi - \frac{e^{-4\phi/(D-2)}}{12} H_{\mu\nu\rho} H^{\mu\nu\rho} \right). \end{aligned} \quad (6)$$

Thus we have obtained the D-dimensional Einstein action including coupling terms with the dilaton and the antisymmetric field. This form for the effective field theory is known as *Einstein frame action*, while (5) is commonly called the *string frame action*. The interpretation of the gravitational aspects of the theory is more clear in the Einstein frame: in the particular case $D = 4$, the variational principle $\delta S = 0$ imposed to this new form of the effective action leads to the equations

$$\nabla_\mu \partial^\mu \phi + c e^\phi - \frac{1}{16} e^{-2\phi} \mathbf{H}^2 = 0, \quad (7)$$

$$\nabla_\delta \mathbf{H}^\delta_{\mu\nu} + 2\nabla_\delta \phi \mathbf{H}^\delta_{\mu\nu} = 0, \quad (8)$$

$$\begin{aligned} R_{\mu\nu} - \frac{1}{2} g_{\mu\nu} R - \frac{c}{2} g_{\mu\nu} e^\phi &= \frac{1}{2} \left(\nabla_\mu \phi \nabla_\nu \phi - \frac{1}{2} g_{\mu\nu} (\nabla \phi)^2 \right) + \\ &\quad + \frac{1}{4} e^{-2\phi} \left(\mathbf{H}_{\mu\rho\delta} \mathbf{H}^{\rho\delta}_\nu - \frac{1}{6} g_{\mu\nu} \mathbf{H}^2 \right), \end{aligned} \quad (9)$$

and the Bianchi identities
$$\nabla_{[\mu} \mathbf{H}_{\mu\rho\delta]} = 0. \tag{10}$$

We can recognize in (9) the Einstein equations with a cosmological function given by $\Lambda = ce^\phi$ and with the energy-momentum tensor of the dilaton and the antisymmetric fields as the source. The relation between the string frame and the Einstein frame formulations can be clarified by considering a solution of the equations (5) in the case of homogeneity and isotropy. Because the metric ds^2 in the string frame is related to the corresponding metric $d\tilde{s}^2$ in the Einstein frame by $ds^2 = e^\phi d\tilde{s}^2$, one can easily translate the results. For example, for the case $c = 0$ a possible solution is a flat cosmology with a metric ds^2 where the scale factor behaves like $a \sim \tau^{1/3}$. By recalling the corresponding evolution of the dilaton with τ, this behavior is translated to the Einstein frame as an evolution of the scale factor $b \sim \tau^{1/2}$, that is, to the same evolution of a radiation-dominated universe (see [27] for a detailed discussion).

2.2 Cosmological models and Hamiltonian formulation

Consider a cosmological model with a finite number of degrees of freedom identified by the coordinates q^i (geometrical and matter degrees of freedom). The Lagrangian form for the action of such a minisuperspace is

$$S[q^i, N] = \int_{\tau_1}^{\tau_2} N \left(\frac{1}{2N^2} G_{ij} \frac{dq^i}{d\tau} \frac{dq^j}{d\tau} - V(q) \right) d\tau \tag{11}$$

where a spatial integration must be understood in the integrand. In (11), G_{ij} is the reduced version of the DeWitt supermetric (see, for example, [6]), V is the potential, which depends on the curvature and the coupling between the fields, and $N(\tau)$ is the lapse function determining the separation between spacelike hypersurfaces in spacetime [6, 29, 45].

The role of constraints and Lagrange multipliers becomes manifest in the Hamiltonian formulation, being this formalism the best suited for the canonical quantization of cosmological models. If we define the canonical momenta as

$$p_i = \frac{1}{N} G_{ij} \frac{dq^j}{d\tau},$$

we obtain the Hamiltonian form of the action

$$S[q^i, p_i, N] = \int_{\tau_1}^{\tau_2} \left(p_i \frac{dq^i}{d\tau} - N\mathcal{H} \right) d\tau, \tag{12}$$

where

$$\mathcal{H} = G^{ij} p_i p_j + V(q). \tag{13}$$

Under arbitrary changes of the coordinates q^i, the momenta p_i and the lapse function N we obtain

$$\delta S = p_i \delta q^i \Big|_{\tau_1}^{\tau_2} + \int_{\tau_1}^{\tau_2} \left[\left(\frac{dq^i}{d\tau} - N \frac{\partial \mathcal{H}}{\partial p_i} \right) \delta p_i - \left(\frac{dp^i}{d\tau} + N \frac{\partial \mathcal{H}}{\partial q_i} \right) \delta q^i - \mathcal{H} \delta N \right] d\tau. \tag{14}$$

Then if we demand the action to be stationary when the coordinates q^i are fixed at the boundaries, we obtain on the classical path the Hamilton canonical equations

$$\frac{dq^i}{d\tau} = N[q^i, \mathcal{H}], \qquad \frac{dp_i}{d\tau} = N[p_i, \mathcal{H}] \qquad (15)$$

and the Hamiltonian constraint

$$\mathcal{H} = 0. \qquad (16)$$

Two features of the dynamics can thus be remarked. The first one is that the presence of the constraint $\mathcal{H} = 0$ restricts possible initial conditions to those lying on the constraint surface. The second is that the evolution of the lapse function N is arbitrary, i.e., it is not determined by the canonical equations; hence, the separation between two successive three-surfaces is arbitrary, which constitutes the minisuperspace version of the many-fingered nature of time of the full theory of gravitation [6, 45].

The field equations yielding from the spacetime action (5) admit homogeneous and isotropic solutions in four dimensions [1, 48, 49, 27]. Such solutions have a metric of the Friedmann–Robertson–Walker form:

$$ds^2 = N(\tau)d\tau^2 - a^2(\tau)\left(\frac{dr^2}{1-kr^2} + r^2 d\theta^2 + r^2 \sin^2\theta d\varphi^2\right), \qquad (17)$$

where a is the scale factor and $k = (-1, 0, 1)$ determines the curvature. For the dilaton ϕ and the field strength $H_{\mu\nu\rho}$ the homogeneity and isotropy requirements lead to

$$\phi = \phi(\tau) \qquad \mathbf{H}_{ijk} = \lambda(\tau)\varepsilon_{ijk} \qquad (18)$$

where ε_{ijk} is the volume form on the constant-time surfaces and λ is a real number. The Bianchi identities (10) imply that λ does not depend on the parameter τ. An important aspect of these cosmological solutions is that they allow for a conceptually new scenario for the early universe, where the standard big bang is replaced by a phase of finite curvature [24, 23].

Let us write down the explicit form of the action for the models to be considered here. If we define $b^2(\tau) \equiv e^{2\Omega(\tau)}$, the Lagrangian form of the Einstein frame action in four dimensions for the case $\lambda = 0$, which corresponds to the two-form field $B_{\mu\nu}$ equal to zero, is given by

$$S = \frac{1}{2}\int_{\tau_1}^{\tau_2} d\tau N e^{3\Omega}\left[-\frac{\dot\Omega^2}{N^2} + \frac{\dot\phi^2}{N^2} - 2ce^\phi + ke^{-2\Omega}\right], \qquad (19)$$

where a dot stands for $d/d\tau$. On the other hand, in the case $k = 0$ (flat universe) we can write

$$S = \frac{1}{2}\int_{\tau_1}^{\tau_2} d\tau N e^{3\Omega}\left[-\frac{\dot\Omega^2}{N^2} + \frac{\dot\phi^2}{N^2} - 2ce^\phi - \lambda^2 e^{-6\Omega-2\phi}\right]. \qquad (20)$$

We have absorbed the factor $(8\pi G_N)^{-1}$ by redefining the fields.

The Hamiltonian form of the Einstein frame action for the models considered reads

$$S = \int_{\tau_1}^{\tau_2} d\tau \left[p_\Omega \dot\Omega + p_\phi \dot\phi - N\mathcal{H}\right]. \qquad (21)$$

For $\lambda = 0$ the Hamiltonian constraint is

$$\mathcal{H} = \frac{1}{2}e^{-3\Omega}\left(-p_\Omega^2 + p_\phi^2 + 2ce^{6\Omega+\phi} - ke^{4\Omega}\right) = 0, \qquad (22)$$

while for $k = 0$ we have

$$\mathcal{H} = \frac{1}{2}e^{-3\Omega}\left(-p_\Omega^2 + p_\phi^2 + 2ce^{6\Omega+\phi} + \lambda^2 e^{-2\phi}\right) = 0. \qquad (23)$$

These two Hamiltonian constraints, or their scaled forms $H \equiv 2e^{3\Omega}\mathcal{H} = 0$ (see Appendix A), will be the starting point for the canonical quantization of the models.

3 Quantization

3.1 Schrödinger and Wheeler–DeWitt equations

The standard procedure for quantizing the minisuperspaces models described by the Hamiltonian constraints (22) and (23) is to turn them into operators and make them act on a wave function (Dirac method). This prescription yields the usual Wheeler–DeWitt equation, which is an hyperbolic equation of second order in all its derivatives. However, the absence of a true time in the classical formalism (which is reflected in the appearance of the Hamiltonian constraint $\mathcal{H} = 0$) makes difficult the interpretation of the resulting wave function in terms of a conserved positive-definite inner product. One knows in fact how to define a conserved positive-definite inner product for the solutions of a Schrödinger equation, and not for the solutions of a Klein–Gordon type equation as the Wheeler–DeWitt equation. Other problems associated with the Wheeler–DeWitt equation are discussed in [40, 41, 42].

However, in certain cases it is possible, by means of the identification of a time variable among the canonical variables, to extract from the time independent Wheeler–DeWitt equation a time dependent Schrödinger equation. In this case a clear probability interpretation can be given to the space of solutions of the Schrödinger equation. As Barbour said [5], in canonical quantum gravity the situation is in a certain sense inverted with respect to ordinary quantum mechanics where the time independent Schrödinger equation $\widehat{h}\psi = E\psi$ is derived from the more general time dependent Schrödinger equation $\widehat{h}\psi = i\frac{\partial \psi}{\partial t}$. In canonical quantum gravity a time dependent Schrödinger equation has to be extracted from a time independent Schrödinger equation with autovalue zero $\widehat{\mathcal{H}}\psi = 0$ (Wheeler–DeWitt equation).

There are mainly two possibilities for defining a Schrödinger equation for a dynamical system with a constraint $\mathcal{H} = 0$ quadratic in all its momenta. The first possibility it to perform a canonical transformation which transforms the quadratic Hamiltonian constraint in a constraint linear in one of its canonical momenta. If a certain positivity condition is satisfied by this linear momentum, its canonically conjugated coordinate can then be defined as the "real" time parameter for the evolution of the system [7, 22, 9].

The other possibility is to factorize the scaled Hamiltonian constraint $H \equiv 2e^{3\Omega}\mathcal{H} = p_0^2 - h^2 = 0$ (with $h^2 = p_\mu p^\mu + V(q^\mu, q^0)$) in two disjoint sheets $H = (p_0 + h)(p_0 - h) = 0$, given by the two signs of the momentum p_0 conjugated to the coordinate q^0 that one wants to identify with time (modulo a sign). Below we shall analyze the conceptual meaning of this factorization. A time dependent Schrödinger equation can then be associated to each factor $p_0 \pm h$. The splitting of the original constraint in two constraints requires the existence of a non vanishing momentum p_0 (a *true* time must not invert its sense of evolution), which may not happen in the original variables describing the model. Indeed, for q^0 to be a right time choice, the condition $[q^0, \mathcal{H}] > 0$ must be fulfilled (see [28] and the Appendix A), leading this condition to the requirement that $p_0 \neq 0$. Besides, for obtaining a proper quantum theory in each sheet, the operator corresponding to the reduced Hamiltonian h must be a self-adjoint operator in order to have a unitary evolution. Given that in the factorization process the reduced Hamiltonian h is obtained by taking a square root $\left(h = \sqrt{p_\mu p^\mu + V(q^\mu, q^0)}\right)$, the associated operator $\hat{h}$ will be self-adjoint only if the term under the square root is positive defined, being this fact an important constraint to the reduction process that we have described. It may also happen that the model can not be reduced in the original canonical variables, being then necessary to perform a canonical transformation in order to properly reduce the system with a self-adjoint Hamiltonian operator $\hat{h}$. In that case it is possible that the selected time variable depends both on the original canonical coordinates and momenta ($t \equiv q^0 = q^0(q^\mu, p_\mu)$), being these kind of time variables called *extrinsic times* [32, 53] (in contrast to the *intrinsic times* which depend only on the canonical coordinates q^μ).

If the reduction process can be successfully performed, then we will have a pair of Hilbert spaces, each one with its corresponding Schrödinger equation of first order in $\partial/\partial q^0$ with a reduced Hamiltonian operator $\hat{h}$. The physical inner product is defined in each space as

$$(\Psi_2|\Psi_1) = \int dq\, \delta(q^0 - \tilde{q}^0) \Psi_2^* \Psi_1. \tag{24}$$

We could say that the Schrödinger quantization preserves the topology of the constraint surface: the splitting of the classical solutions into two disjoint subsets has its quantum version in the splitting of the theory in two Hilbert spaces [10, 46].

3.2 Intrinsic time

We shall begin by considering a generic (scaled) Hamiltonian constraint of the form

$$-p_1^2 + p_2^2 + A e^{(aq^1 + bq^2)} = 0, \tag{25}$$

with $a \neq b$. This constraint admits an intrinsic time because the potential does not vanish for finite values of the coordinates (see Appendix A; for models not fulfilling this condition see the next section). This Hamiltonian constraint corresponds to several dilaton cosmologies, namely, to the cases $\lambda = 0, k = 0, c \neq 0$; $\lambda = 0, k = \pm 1, c = 0$; $\lambda \neq 0, k = 0, c = 0$ (see Ref. [26] for the quantization of the corresponding minisuperspaces within the path integral formulation). It is easy to show that the coordinate change

$$x \equiv \frac{1}{2}(a\Omega + b\phi) \qquad y \equiv \frac{1}{2}(b\Omega + a\phi) \tag{26}$$

leads to the following form of the constraint:

$$H = -p_x^2 + p_y^2 + \zeta e^{2x} = 0, \tag{27}$$

with $sgn(\zeta) = sgn(A/(a^2 - b^2))$. The Wheeler-DeWitt equation corresponding to this constraint is

$$\left(\frac{\partial^2}{\partial x^2} - \frac{\partial^2}{\partial y^2} + \zeta e^{2x}\right)\Psi_\omega(x, y) = 0. \tag{28}$$

The general solution for the case $\zeta > 0$ is

$$\Psi_\omega(x, y) = \left[a_+(\omega)e^{i\omega y} + a_-(\omega)e^{-i\omega y}\right]$$
$$\times \left[b_+(\omega)J_{i\omega}(\sqrt{|\zeta|}e^x) + b_-(\omega)N_{i\omega}(\sqrt{|\zeta|}e^x)\right], \tag{29}$$

with $J_{i\omega}$ and $N_{i\omega}$ the Bessel and Neumann functions of imaginary order respectively. Instead, for $\zeta < 0$, the general solution is

$$\Psi_\omega(x, y) = \left[a_+(\omega)e^{i\omega y} + a_-(\omega)e^{-i\omega y}\right]$$
$$\times \left[b_+(\omega)I_{i\omega}(\sqrt{|\zeta|}e^x) + b_-(\omega)K_{i\omega}(\sqrt{|\zeta|}e^x)\right], \tag{30}$$

where $I_{i\omega}$ and $K_{i\omega}$ are modified Bessel functions. The obtention of these solutions to the Wheeler–DeWitt equation does not finish the quantization process given that we do not know yet which are the proper boundary conditions that we should impose on these spaces of solutions.

Instead of quantizing the model by solving the Wheeler–DeWitt equation, one could also reduce the system by identifying a time variable and solving the corresponding Schrödinger equations. Depending on the sign of the constant ζ in the constraint (27), these models admit as global phase time the coordinates x or y. In the case $\zeta > 0$ the time is $t = \pm x$, so that we can define the reduced Hamiltonian as $h = \sqrt{p_y^2 + \zeta e^{2x}}$, being this reduced Hamiltonian time dependent. If, instead, we have $\zeta < 0$, the time is $t = \pm y$ and the reduced Hamiltonian is $h = \sqrt{p_x^2 - \zeta e^{2x}}$. In the first case, $\zeta > 0$ ($t = \pm x$), the corresponding Schrödinger equations are

$$i\frac{\partial}{\partial x}\Psi(x, y) = \pm\left(-\frac{\partial^2}{\partial y^2} + \zeta e^{2x}\right)^{1/2}\Psi(x, y) \tag{31}$$

In the second case ($\zeta < 0$, $t = \pm y$) the associated Schrödinger equations are

$$i\frac{\partial}{\partial y}\Psi(x, y) = \pm\left(-\frac{\partial^2}{\partial x^2} - \zeta e^{2x}\right)^{1/2}\Psi(x, y). \tag{32}$$

For both $\zeta > 0$ and $\zeta < 0$ we have a pair of Hilbert spaces, each one with its corresponding Schrödinger equation, and a conserved positive-definite inner product allowing for the usual probability interpretation of the wave function (this is analogous to the obtention of two quantum propagators, one for each disjoint theory, in the context of path integral quantization [46, 6]).

It is important to remark that in the present model the time variables which permit to reduce the system ($t = \pm x$ and $t = \pm y$ for the cases $\zeta > 0$ and $\zeta < 0$ respectively) can be selected among the original canonical coordinates, being then unnecessary to perform any kind of canonical transformation. In cases like this one the Schrödinger quantization formalism can be applied directly from the beginning in a straightforward manner.

It is also important to insist on the fact that the cases $\zeta > 0$ and $\zeta < 0$ are not symmetric: if in the case $\zeta < 0$ ($t = \pm y$) the reduced Hamiltonian $h = h(x, p_x)$ is time independent, in the case $\zeta > 0$ ($t = \pm x$) the reduced Hamiltonian $h = h(x, p_y)$ is time dependent.

For the case $\zeta < 0$, the stationary solutions to the Schrödinger equations corresponding to both sheets ($t = \pm y$) are

$$\Psi_\pm^\omega(x, y) = e^{\mp i\omega y}\left[b_+(\omega)J_{i\omega}(\sqrt{|\zeta|}e^x) + b_-(\omega)N_{i\omega}(\sqrt{|\zeta|}e^x)\right] \tag{33}$$

It is clear from this expression that the space of solutions (30) of the Wheeler–DeWitt equation for $\zeta < 0$ is the direct sum of the spaces of solutions of the Schrödinger equations corresponding to each sheet of the Hamiltonian constraint. For the time dependent case ($\zeta > 0, t = \pm x$) the relation between both spaces of solutions is not so clear due to the presence of operator ordering ambiguities (this case will be discussed in Section 3.5).

Another point to be remarked is that as a result of the right definition of time, in both cases the reduced Hamiltonian h are real, so that the evolution operator is self-adjoint and the resulting quantization is unitary. Instead, a wrong choice of time, like for example $t = \pm x$ in the case $\zeta < 0$, leads to a reduced Hamiltonian h which is not real for all allowed values of the variables, and we obtain a non unitary theory. Another remarkable aspect is that a right time is not necessarily associated to the geometrical degrees of freedom, as one could naively expect. In the case $\lambda \neq 0, k = 0, c = 0$ for example, one obtains that the *geometrical* coordinate Ω is the physical clock and that the reduced Hamiltonian involves only the dilaton and the antisymmetric field (the last one, through the constant λ). On the contrary, in the case $\lambda = 0, k = 1, c = 0$ the physical clock is the dilaton, while the reduced Hamiltonian depends only on the geometry (see Section 4.1).

3.3 Extrinsic time

We shall now consider the problem of models not admitting a global time defined only in terms of the coordinates. A good example is given by a nontrivial dilaton cosmology described by the scaled Hamiltonian constraint

$$H = -p_\Omega^2 + p_\phi^2 + 2ce^{6\Omega+\phi} + \lambda^2 e^{-2\phi} = 0, \tag{34}$$

which corresponds to a flat universe with dilaton field ϕ and non vanishing antisymmetric field $B_{\mu\nu}$ coming from the NS-NS sector of effective string theory. This model is not solvable by separating variables, and in the case $c < 0$ the potential can vanish, so that it does not admit an intrinsic time. However, because these cosmologies come from the low energy string theory, which makes sense in the limit $\phi \to -\infty$, then the $e^\phi \equiv V(\phi)$ factor in the first term of the potential verifies $V(\phi) = V'(\phi) \ll 1$, and we can replace ce^ϕ by the constant $\bar{c}$ fulfilling $|\bar{c}| \ll |c|$:

$$H = -p_\Omega^2 + p_\phi^2 + 2\bar{c}e^{6\Omega} + \lambda^2 e^{-2\phi} = 0. \tag{35}$$

Following Ref. [37], we shall start in a naive way by quantizing by means of a Wheeler–DeWitt equation the model described by the Hamiltonian constraint (35). However, for this model we can also perform a canonical transformation which permits to obtain a Hamiltonian constraint that can be factorized in a two-sheet constraint of the form $H = (p_0 + h)(p_0 - h) = 0$ [12]. We will then be able to quantize the model also by using the Schrödinger equations corresponding to each sheet of the factorized Hamiltonian constraint. It is important to remark that this Schrödinger quantization is not trivial in the sense that a non trivial canonical transformation is needed. If this canonical transformation were not known, we should face the problem of imposing boundary conditions on the space of solutions of the Wheeler–DeWitt equation. Given that this model can be quantized by both methods, it is of great value in order to understand how to impose boundary conditions to the solutions of the Wheeler–DeWitt equation for cases in which a global time is not known.

The Wheeler–DeWitt equation associated to the constraint (35) is

$$\left(\frac{\partial^2}{\partial \Omega^2} - \frac{\partial^2}{\partial \phi^2} + 2\bar{c}e^{6\Omega} + \lambda^2 e^{-2\phi}\right)\Psi(\Omega, \phi) = 0, \qquad (36)$$

and its solutions are

$$\Psi_\omega(\Omega, \phi) = \left[a(\omega)I_{i\omega}\left(|\lambda|e^{-\phi}\right) + b(\omega)K_{i\omega}\left(|\lambda|e^{-\phi}\right)\right] \\ \times \left[c(\omega)I_{i\omega/2}\left(\sqrt{|2\bar{c}|}e^{3\Omega}\right) + d(\omega)K_{i\omega/2}\left(\sqrt{|2\bar{c}|}e^{3\Omega}\right)\right], \qquad (37)$$

where I_ν and K_ν are modified Bessel functions. In principle one does not know which are the proper boundary conditions that should be imposed on these solutions. If one was to proceed in a naive way without any consideration about time, then both contributions including the functions I_ν should be discarded, because they diverge in what would be commonly understood as a region classically forbidden by the behavior of the exponential terms in the potential. In fact, this has been the choice in the case of the Taub universe in Ref. [37]. However, as we shall show, the functions $I_{i\omega}(|\lambda|e^{-\phi})$ should not be discarded, because in the picture including a globally right notion of time, the dilaton ϕ is associated to the physical clock.

This can be clearly realized by performing the canonical transformation introduced for the Taub universe in Ref. [12] in order to obtain a constraint with only one term in the potential. This is achieved by introducing the generating function of the first kind

$$\Phi_1(\phi, s) = \pm|\lambda|e^{-\phi}\sinh s. \qquad (38)$$

The new canonical variables are then given by

$$s = \pm \text{arcsinh}\left(\frac{p_\phi e^\phi}{|\lambda|}\right) \\ p_s = \pm|\lambda|e^{-\phi}\cosh s. \qquad (39)$$

With this canonical transformation the resulting form for the Hamiltonian constraint in the limit $V(\phi) = V'(\phi) \ll 1$ is

$$H = -p_\Omega^2 + p_s^2 + 2\bar{c}e^{6\Omega} = 0, \qquad (40)$$

and we can apply our procedure starting from this constraint. In the case $\bar{c} < 0$, the momentum p_s does not vanish and the time is $t = \pm s$. According to the definition of the new variable s the time $t = \pm s$ is a function of both p_ϕ and ϕ, being then an *extrinsic time*. Once again, observe that, differing from what is sometimes considered the "natural" choice, the physical clock is not associated to the metric, but to the matter field (see below). The constraint (40) can be written as

$$H = \left(p_s - \sqrt{p_\Omega^2 - 2\bar{c}e^{6\Omega}}\right)\left(p_s + \sqrt{p_\Omega^2 - 2\bar{c}e^{6\Omega}}\right) = 0 \quad (41)$$

(see Section 3.5). If this constraint equation is satisfied by demanding $p_s = h = \sqrt{p_\Omega^2 + |2\bar{c}|e^{6\Omega}}$, then this choice implies (taking into account that $p_t = -h$) that one has selected the variable $-s$ to play the role of the physical clock ($t = -s$ and $p_t = -p_s$). If on the contrary one demands $p_s = -h = -\sqrt{p_\Omega^2 + |2\bar{c}|e^{6\Omega}}$, then one is choosing s as the physical clock ($t = s$ and $p_t = p_s$). In both cases the positive definite reduced Hamiltonian h does not depend on the time variable $t = \pm s$. As we will explain later, the fact that the reduced Hamiltonian h is time independent assures the equivalence at the quantum level of the factorized constraint (41) and the original one given by (40). The corresponding Schrödinger equations associated to the choices $t = s$ and $t = -s$ are respectively

$$i\frac{\partial}{\partial s}\Psi_+(\Omega, s) = \left(-\frac{\partial^2}{\partial \Omega^2} + |2\bar{c}|e^{6\Omega}\right)^{1/2}\Psi_+(\Omega, s). \quad (42)$$

$$-i\frac{\partial}{\partial s}\Psi_-(\Omega, s) = \left(-\frac{\partial^2}{\partial \Omega^2} + |2\bar{c}|e^{6\Omega}\right)^{1/2}\Psi_-(\Omega, s). \quad (43)$$

By $\Psi_\pm$ we have noted the solutions corresponding to the sheets $K_\pm = p_s \pm h = 0$. Note that $\Psi_+ = (\Psi_-)^*$ (this relation between both spaces of solutions will be discussed later). Given that the reduced Hamiltonian $h = \sqrt{p_\Omega^2 - 2\bar{c}e^{6\Omega}}$ does not depend on time, one can propose a solution of the general form $\Psi_E(\Omega, t) = \varphi_E(\Omega)e^{-iEt}$, where $\varphi_E(\Omega)$ satisfies the time independent Schrödinger equation $\hat{h}(\Omega)\varphi_E(\Omega) = E\varphi_E(\Omega)$. Given that $\hat{h}$ is a square root operator, one has to use the spectral theorem and solve the derived equation $\hat{h}^2(\Omega)\varphi_E(\Omega) = E^2\varphi_E(\Omega)$ [8]. The general solution to this last equation is

$$\varphi_\omega(\Omega) = \left[c(\omega)I_{i\omega/2}\left(\sqrt{|2\bar{c}|}e^{3\Omega}\right) + d(\omega)K_{i\omega/2}\left(\sqrt{|2\bar{c}|}e^{3\Omega}\right)\right] \quad (44)$$

The functions $I_{i\omega/2}\left(\sqrt{|2\bar{c}|}e^{3\Omega}\right)$ have to be discarded because they diverge in the classical forbidden zone $\Omega \to \infty$. The stationary solutions of the Schrödinger equations (42) and (43) are then of the form

$$\Psi_\pm^\omega(\Omega, s) = d(\omega)K_{i\omega/2}\left(\sqrt{|2\bar{c}|}e^{3\Omega}\right)e^{\mp i\omega s} \quad (45)$$

It is important to remark that the equation $\hat{h}^2(\Omega)\varphi_E(\Omega) = E^2\varphi_E(\Omega)$ is the same equation that one obtains from the Wheeler-DeWitt equation associated to the constraint (40) by separating variables (i.e., by proposing solutions of the form $\Psi(\Omega, s) = \varphi(\Omega)\phi(s)$). This means that the equation that one has to solve *in any case* is a second order Wheeler-DeWitt equation of the hyperbolic type. The reduction process changes only the variables in which this equation will be solved. The only difference is that in the new set of variables it is clear which are the boundary conditions to be imposed.

3.4 Relation between solutions of the Schrödinger and the Wheeler-DeWitt equations

As we have shown in the preceding examples, there are mainly two quantization schemes. Being the Wheeler-DeWitt equation an hyperbolic equation quadratic in all its derivatives, its space of solutions is twice the space of solutions of the parabolic Schrödinger equation linear in $\frac{\partial}{\partial t}$. If the reduced Hamiltonian h does not depend on time, the space of solutions of the Wheeler-DeWitt equation $\widehat{H}\Psi = 0$ will be simply the direct sum of the spaces of solutions of the Schrödinger equations corresponding to each sheet of the Hamiltonian constraint:

$$Ker\widehat{H} = Ker\widehat{K_+} \oplus Ker\widehat{K_-}, \tag{46}$$

where $\widehat{K_+}$ and $\widehat{K_-}$ are the operators corresponding to the factors $p_0 + h$ and $p_0 - h$ respectively.

It could happen nevertheless that, even if the reduced Hamiltonian h is time independent, the correspondence between the solutions of the Wheeler-DeWitt equation and the solutions of the Schrödinger equation is not so direct. For the dilaton cosmology described by the Hamiltonian constraint (34), the passage from the Hamiltonian constraint in its original form to the factorized form (41) is mediated by the canonical transformation given by (39). The solutions (37) of the Wheeler-DeWitt equation (36) and the solutions of the Schrödinger equations (42) and (43) are not expressed in terms of the same variables ((Ω, ϕ) and (Ω, s) respectively). In Ref. [12], this situation was analyzed for the Taub model and a certain criterium was proposed for fixing boundary conditions to the Wheeler-DeWitt equation for those cases where we do not know how to reduce the system. We will now describe such a proposal.

In Ref. [12] it was proposed that the solutions of the Wheeler-DeWitt equation (36) can be related with the solutions of the Schrödinger equations (42) and (43) by means of a quantum version of the classical canonical transformation (39) used to reduce the system. Under certain conditions two quantum systems whose Hamiltonians are canonically equivalent at the classical level, have as quantum states wave functions which can be related by means of the so called "quantum canonical transformations". This quantum version of the classical canonical transformations can be understood as a generalization of the Fourier Transform. It is in fact possible to consider the Fourier transform as the quantum version of the classical canonical transformation which interchanges coordinates and momenta. The generating function of such a canonical transformation is $F_1(q, Q) = Qq$ and the equations defining the transformation are

$$p = \frac{\partial F_1}{\partial q} = Q, \qquad P = -\frac{\partial F_1}{\partial Q} = -q.$$

The Fourier transform

$$\Psi(q) = N \int dp\, e^{ipq} \Phi(p) \tag{47}$$

can then be rewritten as

$$\Psi(q) = N \int dQ \, e^{iF_1(q,Q)} \Phi(Q), \qquad (48)$$

The inverse of this transformation could be expressed as

$$\Phi(Q) = N \int dq \left| \frac{\partial^2 F_1(q,Q)}{\partial q \partial Q} \right| e^{-iF_1(q,Q)} \Psi(q). \qquad (49)$$

It is then natural to ask if these expressions remain valid for a canonical transformation given by a general generating function $F_1(q,Q)$. If this were the case, one would have a Generalized Fourier Transform between the quantum representations associated to systems canonically equivalent at the classical level. In general this is not the case: certain conditions must be fulfilled in order to have this kind of Generalized Fourier Transforms. In [25] it was shown that the expression (48) is valid if the following condition is satisfied

$$H_q\left(-i\frac{\partial}{\partial q}, q\right) e^{iF_1(q,Q)} = H_Q\left(i\frac{\partial}{\partial Q}, Q\right) e^{iF_1(q,Q)}, \qquad (50)$$

where certain boundary conditions in the integration limits are also assumed. If the canonical transformation is defined by means of other kind of generating function (F_2, F_3 or F_4) analogous expressions can be used.

In particular, the canonical transformation (39) used to pass from the Hamiltonian constraint (34) to the factorized Hamiltonian constraint (41) effectively satisfies these requirements. The solutions of the Schrödinger equations (42) and (43) can then be related to the solutions of the Wheeler-DeWitt equation (36) by means of the corresponding Generalized Fourier Transform.

By means of this formalism we can now analyze the proper boundary conditions that should be imposed on the space of solutions of the Wheeler-DeWitt equation (36). Firstly, one should notice that the dependance on Ω is the same for both representations ((Ω, ϕ) and (Ω, s)). This means that one can apply, on the factor depending on Ω in the solutions (37) of the Wheeler-DeWitt equation, the same boundary conditions that we have previously imposed on the solutions of the Schrödinger equations. In fact, given our choice of the physical clock as a function of both ϕ and p_ϕ (see (39)), the variable Ω is an authentical dynamical variable. In this way one can discard in the solution (37) the functions $I_{i\omega/2}\left(\sqrt{|2c|}e^{3\Omega}\right)$ because they diverge in the classical forbidden zone $\Omega \to \infty$. Next we have to impose boundary conditions on the factor which depends on ϕ in (37), i.e., we have to decide if we will discard the functions $I_{i\omega}\left(|\lambda|e^{-\phi}\right)$ or the functions $K_{i\omega}\left(|\lambda|e^{-\phi}\right)$. The criterium that we will apply is that the physical solutions will be those whose transformed functions (via the generalized Fourier transform) coincide with the factors $e^{-i\omega s}$ or $e^{i\omega s}$ in the solutions $\Psi\pm$ (45) of the Schrödinger equations (42) and (43) for the clock choices $t = s$ or $t = -s$ respectively. One could suppose that the correct solutions will be those which go to zero in the classical forbidden zone $\phi \to -\infty$, i.e., the functions $K_{i\omega}\left(|\lambda|e^{-\phi}\right)$. But if we transform this functions we have to conclude that they have to be discarded given that they correspond to a linear combination of the form $ae^{i\omega s} + be^{-i\omega s}$. If on the contrary we apply the generalized Fourier transform to the functions $I_{\pm i\omega}\left(|\lambda|e^{-\phi}\right)$ we obtain the correspondence $I_{\pm i\omega}\left(|\lambda|e^{-\phi}\right) \leftrightarrow e^{\mp i\omega s}$. The functions $I_{i\omega}\left(|\lambda|e^{-\phi}\right)$ and $I_{-i\omega}\left(|\lambda|e^{-\phi}\right)$ do represent then the positive energy states corresponding to the clock choices $t = s$ and $t = -s$ respectively. If we make the choice $t = s$ the solution to the Wheeler-DeWitt equation (36) is then

$$\Psi_\omega(\Omega, \phi) = \tilde{a}(\omega) I_{i\omega}\left(|\lambda|e^{-\phi}\right) K_{i\omega/2}\left(\sqrt{|2\bar{c}|}e^{3\Omega}\right) \qquad (51)$$

The fact that the correct solutions are not those which go to zero in the classical forbidden zone $\phi \to -\infty$ does not pose a problem given that the variable ϕ is not a dynamical variable, but the variable associated with the physical clock that we have chosen. This fact constitutes an important difference with the immediate result that we would obtain by following Ref. [37].

It could be argued that the real problem is to find proper boundary conditions for the Wheeler-DeWitt equation when one does not know how to reduce the system. If one has a reduced system with the corresponding Schrödinger equation as in the Taub case, it is no more necessary to go back to the Wheeler-DeWitt "representation". However, given that for the Taub model both spaces of solutions are known, its analysis is of great utility in order to propose general boundary conditions to the Wheeler-DeWitt equation, even for those cases for which one does not know how to separate a physical clock. The main requirement to impose on the general boundary conditions that we are looking for is that, for those cases for which one knows how to reduce the model (as the Taub universe), they have to select the same quantum states imposed by the quantization of the reduced system. In Ref. [12] certain steps were given in this direction. We will now describe these results.

Let us suppose as a first simplification that we have the following scaled Hamiltonian constraint

$$H = p_0^2 + V\left(q^0\right) - h^2\left(q^\mu, p_\mu\right) \qquad (52)$$

with $V(q^0) > 0$. In other words, let us suppose that there is not a non-minimal coupling between a certain variable q^0 and the rest of the canonical variables q^μ. Given this form of the Hamiltonian constraint, the solutions of the corresponding Wheeler-DeWitt equation will have the form $\Psi(q^0, q^\mu) = \Theta(q^0) \Phi(q^\mu)$. If the potential $V(q^0)$ were identically zero, the variable q^0 would be a physical clock. We could suppose that the true physical clock will be a certain function of q^0 which, in the region where $V(q^0) \to 0$, coincides with q^0. We can expect then that the solutions $\Psi(q^0, q^\mu)$ of the Wheeler-DeWitt equation corresponding to (52) would tend, in the region where $V(q^0) \to 0$, to wave functions of the form $e^{-iEq^0}\Phi(q^\mu)$. The boundary conditions to be imposed on the space of solutions of the Wheeler-DeWitt equation is that the physical solutions will be those functions whose asymptotic expressions in the region where $V(q^0) \to 0$ are of the form $e^{-iEq^0}\Phi(q^\mu)$ (with $\Phi(q^\mu)$ going to zero in the classical forbidden zone associated to q^μ). If we do not know which is the physical clock for reducing the system, but we know that there is a variable q^0 which is a physical clock *in a certain limited region* with a time independent reduced Hamiltonian h, then we can imposed as boundary conditions that the physical solutions have to tend in that region to functions with a factor e^{-iEq^0}.

If we now apply this criterium to the solutions of the Wheeler-DeWitt equation (36) corresponding to the Taub model, we will select the correct quantum states, i.e. the quantum states (51) selected by the reduction of the model. The Hamiltonian constraint (35) of the Taub model has in fact the form (52). In the region where the potential term $\lambda^2 e^{-2\phi}$ goes to zero, i.e., in the region $\phi \to \infty$, the variable ϕ is a physical clock. Following the proposed criterium, we have to select those functions which tend to functions with a factor $e^{-iE\phi}$ in the region $\phi \to \infty$. If we consider the asymptotic expressions of the functions $K_{i\omega}\left(|\lambda|e^{-\phi}\right)$ and $I_{\pm i\omega}\left(|\lambda|e^{-\phi}\right)$ in that limit, we will find that the former tend to a linear combination of the form $ae^{iE\phi} + be^{-iE\phi}$, while only the latter tend to functions of the form $e^{\pm iE\phi}$.

3.5 Time dependent reduced Hamiltonians

A serious problem for the understanding of these two quantization formalisms and its relations appears when the reduced Hamiltonian h is time dependent. While the constraints $H = p_0^2 - h^2 = 0$ and $H = (p_0 + h)(p_0 - h) = 0$ are classically equivalent, at the quantum level this equivalence is no more fulfilled if the reduced Hamiltonian h depends on the variable chosen as physical clock. In fact a wave function in the kernel of the operators $\widehat{K}_+$ and $\widehat{K}_-$ corresponding to the factors $(p_0 + h)$ and $(p_0 - h)$ respectively (i.e., a solution of the Schrödinger equation), is not necessarily annihilated by the operator $\widehat{H}$ associated with the Wheeler-DeWitt equation. If the reduced Hamiltonian h is time dependent the product of the Schrödinger operators $\widehat{K}_+$ and $\widehat{K}_-$ is not equal to the Wheeler-DeWitt operator $\widehat{H}$. The two possible products of the Schrödinger operators are

$$\widehat{K}_\pm \widehat{K}_\mp = -\frac{\partial^2}{\partial t^2} - \hat{h}^2 \mp \left[-i\frac{\partial}{\partial t}, \hat{h}\right] \tag{53}$$

If $h \neq h(t)$ (i.e., $[p_t, h] = 0$), then $\widehat{K}_+\widehat{K}_- = \widehat{K}_-\widehat{K}_+ = \widehat{H} = -\frac{\partial^2}{\partial t^2} - \hat{h}^2$. If $h = h(t)$, then the Wheeler-DeWitt operator $\widehat{H}$ is equal to

$$\widehat{H} = \frac{1}{2}\left(\widehat{K}_+\widehat{K}_- + \widehat{K}_-\widehat{K}_+\right) \tag{54}$$

It is then clear that the solutions to the Schrödinger equations $\widehat{K}_\pm \Psi_\pm = 0$ are not necessarily solutions of the Wheeler-DeWitt equation $\widehat{H}\Psi = 0$ when the reduced Hamiltonian h is time dependent [8]. As it was explained in [8], if the reduced Hamiltonian h is time dependent the solution to the Schrödinger equation $i\frac{\partial \Psi(x,t)}{\partial t} = \hat{h}(x,p,t)\Psi(x,t)$ takes the form

$$\Psi(x,t) = T\left[e^{-i\int_{t_0}^t \hat{h}(t')dt'}\right]\Psi(x,t_0), \tag{55}$$

where T is the time-ordering operator. If the condition

$$[h(t), h(t')] = 0 \tag{56}$$

is satisfied, the expression (55) gives

$$\Psi(x,t) = e^{-i\int_{t_0}^t \hat{h}(t')dt'}\Psi(x,t_0). \tag{57}$$

The condition (56) implies also that there exists a conserved complete set of eigenstates of the Hamiltonian operator $\hat{h}(t)$, i.e., a set of basis states which are eigenstates of $\hat{h}$ at all times. If $\Psi_E(x)$ is an eigenstate of $\hat{h}(t_0)$, then $\Psi_E(x)$ will be an eigenstate of $\hat{h}(t)$ for all times, i.e., there will be a function $E(t)$ such that

$$\hat{h}(t)\Psi_E(x) = E(t)\Psi_E(x). \tag{58}$$

The time evolution of such a state is given by

$$\Psi_E(x,t) = e^{-i\int_{t_0}^{t} E(t')dt'} \Psi_E(x,t_0) \tag{59}$$

An example of a system with a time dependent reduced Hamiltonian $h(t)$ is provided by the dilatonic cosmological model corresponding to the Hamiltonian constraint (27) for $\zeta > 0$. The factorized form of this constraint is

$$H = \left(-p_x + \sqrt{p_y^2 + \zeta e^{2x}}\right)\left(p_x + \sqrt{p_y^2 + \zeta e^{2x}}\right) = 0, \tag{60}$$

being the potential time dependent for $t = \pm x$. Therefore, though at the classical level this product is equivalent to the constraint (27), in its quantum version both constraints differ in terms associated to commutators between $\hat{p}_x$ and the potential ζe^{2x}. The general form of these commutators is $\left[\sqrt{\sum(\hat{p}_\mu)^2 + V(\hat{q}^i)}, \hat{p}_0\right]$ (where $\mu \neq 0$, and V stands for the potential in the scaled Hamiltonian constraint H). Hence, depending on which of the two classically equivalent constraints we start from, we obtain different quantum theories. Observe that this problem appears in the case for which the Wheeler-DeWitt equation leads to a result in which the usual identification of positive and negative-energy solutions is not apparent, at least in the standard form of exponentials of the form $e^{i\omega t}$. In this sense, recall the difference between the solutions (29) and (30).

The central obstruction for the existence of a trivial correspondence between the Wheeler-DeWitt and Schrödinger solutions for minisuperspaces is then a constraint with a time-dependent potential. For the class of models of Section 3.2, a coordinate choice avoiding the decision between inequivalent quantum theories can be introduced [46]. Consider the constraint (25) and define

$$u = \alpha e^{\frac{1}{2}(aq^1+bq^2)}\cosh\left(\frac{bq^1+aq^2}{2}\right), \quad v = \alpha e^{\frac{1}{2}(aq^1+bq^2)}\sinh\left(\frac{bq^1+aq^2}{2}\right), \tag{61}$$

with $\alpha = \sqrt{|A|}$. These coordinates allow to write the scaled constraint in the equivalent form

$$H = -p_u^2 + p_v^2 + \eta m^2 = 0, \tag{62}$$

with $\eta = sgn(A)$ and $m^2 = 4/|a^2 - b^2|$. Given that the commutators do not appear now, the Wheeler-DeWitt equation is equivalent to the corresponding Schrödinger equations, being the physical clock the coordinates u or v depending on η: for $\eta = 1$ we have $p_u \neq 0$ and $t = \pm u$, while for $\eta = -1$ we have $p_v \neq 0$ and $t = \pm v$.

Of course, such a solution can be applied only for a limited number of minisuperspace models. We believe that the case of a time dependent reduced Hamiltonian $h = h(t)$ is the general case. In fact it seems highly improbable that one could find a degree of freedom for playing the role of a physical clock which would not be coupled to the others degrees of freedom (besides the minimal coupling given by the Hamiltonian constraint). In other words, we would not expect to find a "free clock", i.e. a degree of freedom without a non-minimal coupling with the others degrees of freedom. The reduced system will be then in general an open system, i.e., a system which interacts with the clock [17]. If the reduced Hamiltonian h is time dependent, the energy of the reduced system will change. This fact could seem bizarre if one takes into account that in quantum cosmology we are considering the whole universe. But in that case the physical clock will have its own energy, being the change in the energy of the reduced system a consequence of its interaction with the clock (which is nothing but a particular interacting degree of freedom). This change in the "reduced" energy of the system will correspond then to a variation in the momentum of the physical clock ($p_t = -h$): if the clock changes its rate of evolution, the total energy of the others degrees of freedom will change.

4 General aspects of the problem of time

4.1 Parametrized system formalism

A general framework for understanding the meaning of the Hamiltonian constraint $H = 0$ is provided by the formalism of parametrized systems (for details, see [31]; for recent developments, see for example [21, 47, 12, 11, 2, 50, 3]). In this formalism the "real" time t is added to the canonical variables of an ordinary dynamical system, being the increased set of variables left as a function of a physically irrelevant parameter τ. The ordinary action of a dynamical system

$$S[q^\mu, p_\mu] = \int_{t_1}^{t_2} p_\mu dq^\mu - h(q^\mu, p_\mu) \, dt, \quad \mu = 1, ..., n \qquad (63)$$

can then be converted into a parametrized action by means of the definition of a new pair of canonical variables $\{q^0 = t, p_0 = -h\}$ associated to the time t and the Hamiltonian h. The extended set of variables are left as functions of the physically irrelevant parameter τ. The set $\{q^0, q^\mu, p_0, p_\mu\}$ can be varied independently, provided that the definition of p_0 is incorporated into the action as a constraint

$$H = p_0 + h = 0, \qquad (64)$$

with the corresponding Lagrange multiplier N (lapse function). In this way one obtains the following action

$$S\left[q^i(\tau), p_i(\tau), N(\tau)\right] = \int_{t_1}^{t_2} \left(p_i \frac{dq^i}{d\tau} - NH\right) d\tau, \quad i = 0, ..., n \qquad (65)$$

The presence of the Lagrange multiplier $N(\tau)$ means that the dynamics remains ambiguous in the irrelevant parameter τ (one could say that it has no sense to speak about dynamics until the hidden time is recovered). In the extended configuration space $\{q^0 = t, q^\mu\}$ the solutions of the equations of motions are static "curves" without a preferred parametrization. The theory is then invariant under reparametrizations of the physical irrelevant parameter τ. We could say that it has no sense to speak about the speed of the motion of the system through its extended configuration space. These curves have nevertheless a preferred sense: they unfold in the increasing directions of $q^0 = t$.

In this way any dynamical system can be formulated in the framework of parametrized systems. In the process of parametrization, the real time is "disguised" as a dynamical variable. This disguised time can be nevertheless easily recognized due to the linearity of its conjugated momentum in the constraint (64).

It has been intended to use the formalism of parametrized system as a model for understanding the canonical structure of General Relativity. Having the theory a Hamiltonian constraint of the form $H = 0$, one could suppose that General Relativity is an ordinary dynamical theory whose action is presented in a parametrized form. If that were the case, one could try to reduce the system by finding the disguised "real" time and formulating the theory as an ordinary dynamical system of the form (63), whose evolution takes place in that "real" time variable. There has been many proposals for that hidden time variable but none of them could circumvent the different problems which appear in the reduction process [33, 34]. Besides the lack of a universal "real" time variable, there is still another important objection against this interpretation of the canonical structure of General Relativity: the constraint (64) is linear in the momentum p_0, while the Hamiltonian constraint of General Relativity is quadratic in all its momenta.

In [11] a different interpretation was proposed in order to circumvent this problem. Here we will follow this new conceptual framework. We will consider that the supposition of a privileged "real" time hidden among the canonical variables is an unfounded supposition. We consider that one of the most fundamental properties of General Relativity is that its solutions do not represent an *evolution in time* of certain dynamical variables, but that it is a theory which selects certain relative (not dynamical) configurations of its canonical variables which, under certain conditions, can be considered as dynamical evolutions if proper *physical clocks* are selected. In fact we can never observe the evolution of the degrees of freedom along a "newtonian" time flow like $q^1(t)$ and $q^2(t)$ but rather the evolution of certain variables relative to the change of another variable, i.e., something like $q^2(q^1)$. Following this interpretation, there would not be such a thing as Time, but only physical degrees of freedom which, under certain conditions, can be used as physical clocks. In this relational approach we cannot say that reducing the system means to find the hidden "real" time: in order to reduce the model we have only to select a monotonously increasing canonical variable as a physical clock. It is thus only possible to speak about physical clocks, i.e., degrees of freedom which, under certain conditions, can play the role of evolution parameters for the others degrees of freedom. In that case it is not important if the physical clock is a geometrical degree of freedom or a degree of freedom associated to the matter fields (see, for example, [9]); in fact, among the examples of sections 3.2 and 3.3, we identified models with the dilaton field playing the role of physical clock. On the contrary, if we were looking for the "real" time, we could expect that this hidden time would be a geometrical degree of freedom.

The works [30, 21, 43, 44, 18] have shown that the general covariance of General Relativity is not so different from the gauge symmetry of an ordinary gauge theory (see Appendix B). In fact, we could consider General Relativity as a particular case of a gauge theory. In general, in gauge theories the gauge fixing is not interpreted as the discovery of the "real" degrees of freedom, but as a particular choice without any "god given" privilege.

In the framework of gravitation the election of a physical clock acts in fact as a partial gauge fixing of the theory (election of the Lagrange multiplier $N(\tau)$). If we follow the ordinary interpretation of gauge theories this election should not be interpreted as the discovery of the real hidden time, but as a choice of a particular reference system for measuring the evolution of the system.

We could now ask which is the proper formalism for implementing such an interpretation. In [11] a certain modification of the parametrized formalism was proposed in order to explore the consequences of this new conceptual framework.

The first point that one should take into account is that, as there is not a privileged time variable t and a privileged conjugated momentum $p_t = -h$, all the momenta must appear on an equal footing in the Hamiltonian constraint (as effectively happens in General Relativity). This means that there should not be a preferred momentum appearing linearly as in (64). It is then a consequence of the new interpretation the necessity of finding a formalism where all the momenta appear quadratically in the Hamiltonian constraint.

There is still a second argument for implementing a formalism of parametrized systems with a constraint quadratic in all its momenta. The solutions of the theory are static trajectories in a certain configuration space. In order to describe these static trajectories as dynamical evolutions, we have to choose a certain variable to play the role of time. The main difference with the original formalism of parametrized systems is that these trajectories do not carry a preferred sense of evolution. We can choose to unfold them in both directions. This ambiguity can be cast into the parametrized system formalism by means of a modified Hamiltonian constraint. If there is not a privileged time the solutions are necessarily static trajectories, i.e., relative configurations among the different variables, for example $q^1(q^0)$. If one wants now to select a physical clock, for example q^0 (we are supposing that q^0 is a monotonously increasing function along the trajectory) there is still an ambiguity: one still has to choose in which direction the trajectory will be unfold. This means that one can choose $t = q^0$ or $t = -q^0$. The static trajectory does not privilege any direction and then both kinds of solutions must appear in the reduced formalism. These two options correspond to the constraints $K_+ = p_0 + h = 0$ and $K_- = p_0 - h = 0$. Both possibilities can be incorporated into the formalism if we use a Hamiltonian constraint of the form $H = (p_0 + h)(p_0 - h) = p_0^2 - h^2 = 0$. In order to make this factorization the original Hamiltonian constraint must be quadratic in the momentum conjugated to q^0. The existence of two sheets in the Hamiltonian constraint, far from being an unnecessary redundancy, acquires in this way a precise meaning related to the necessity of having a parametrized formalism which does not privilege any sense of evolution. Each sheet of the Hamiltonian constraint $H = 0$ is then associated to each possible choice of the direction in which the trajectory can be unfold. The resulting action is

$$S\left[q^i(\tau), p_i(\tau), N(\tau)\right] = \int p_\mu dq^\mu + p_0 dq^0 - N\left(p_0^2 - h^2\right) d\tau \qquad (66)$$

The constraint $H = 0$ is fulfilled if one of the factors vanishes on the constraint surface. To choose which factor is null is equivalent to choose which direction of q^0 is the increasing direction of time. This system can then be reduced by the clock choices $t = q^0$ or $t = -q^0$, corresponding these choices to the sheets $p_0 + h = 0$ and $p_0 - h = 0$ respectively. If $t = q^0$, then $p_t = p_0 = -h$; if $t = -q^0$, then $p_t = -p_0 = -h$. In both cases $p_t = -h$

with $h > 0$. Having fixed a factor to zero in the Hamiltonian constraint the other one will have, on the constraint surface, a definite sign, so being possible to rescale the Hamiltonian by this factor. If for example we fix $K_+ = p_0 + h = 0$ ($t = q^0$ as physical clock), we will have $p_0 = -h < 0$ and then the other factor will be $K_- = p_0 - h = -2h < 0$.

In this way, both the necessity of having a Hamiltonian constraint where all the momenta appear on an equal footing (quadratically) and the necessity of having the possibility of choosing the direction of unfolding of any static trajectory in both directions, conduce naturally to a modification of the parametrized formalism. We believe that this *quadratic parametrized formalism* is the proper formalism for understanding the canonical structure of General Relativity.

4.2 Motion-reversal and clock-reversal transformations

Without being precise, one could say each choice in the sense of evolution of the physical clock ($t = q^0$ or $t = -q^0$) corresponds to a kind of "time reversal" of the other one. But by choosing a physical clock, i.e., by choosing a sheet of the Hamiltonian constraint $H = (p_0 + h)(p_0 - h) = 0$, one obtains an ordinary dynamical system. It is well known that an ordinary classical or quantum system possess a certain symmetry under "time reversals". If each sheet posses a symmetry under "time reversals", the passage from one sheet to the other one can not be identified with the same time reversal operation. In [11] this situation was clarified. The result of this analysis is that there are two kind of "time reversals" operations which have to be carefully differentiated.

The first one is the ordinary "time reversal" operation of classical and quantum mechanics. In fact this operation does not correspond to an inversion in the direction of time, but to an inverted movement which unfolds in the same direction of time than the original solution. In [11] this operation was called *motion-reversal transformation*. Given a classical trajectory $\{q(q_0, p_0, t_0, t), p(q_0, p_0, t_0, t)\}$ which unfolds from $\{q_0, p_0\}$ at time t_0 to $\{q_f, p_f\}$ at time t_f, there exists a related trajectory which is also a solution of the same Hamilton equations. This inverted trajectory is

$$q^{mr}(q_0^{mr} = q_f, p_0^{mr} = -p_f, t_0, t) = q(q_f, -p_f, t_0, t), \qquad (67)$$
$$p^{mr}(q_0^{mr} = q_f, p_0^{mr} = -p_f, t_0, t) = p(q_f, -p_f, t_0, t),$$

and exists provided that the Hamiltonian h is quadratic in p and does not depend on t. This motion-reversed solution $\{q^{mr}, p^{mr}\}$ unfolds from $\{q_f, p_f\}$ at time t_0 to $\{q_0, p_0\}$ at time t_f. It is important to emphasize that this motion-reversed solution is a solution of the *same* Hamilton equations. This solution starts at the same time t_0 than the original one and unfolds in the same direction of time, but with initial conditions which have been inverted with respect to the original trajectory: the new trajectory starts with an inverted momentum from the point where the original one ends. This operation does not correspond to the operation of "playing the film backwards". In order to see this it is necessary to realize that, being the clock a dynamical variable, it has to be included in the hypothetical film. If we play the film backwards, we will see the clock running backwards. However, we have just said that the motion-reversal transformation of a given solution unfolds in the same direction of time, i.e., the clock must continue to run forward. In other words, the motion-reversed solution corresponds, not to play the same film backwards but to play *another* film in which all behaves as running backwards, but the clock is still running forward. The motion-reversal transformation reverses then all variables, *but the variable identified with physical clock*. In the extended configuration space which includes the physical clock the result of a motion-reversal transformation it is not the same curve (i.e., the same film) unfolded in the opposite direction, but *another* curve which unfolds in the same direction of time.

On the other hand we have a symmetry associated to the passage from one sheet of the Hamiltonian constraint $H = (p_0 + h)(p_0 - h) = 0$ to the other one. This operation involves *also* a change in the direction of unfolding of the variable used as physical clock. This operation was called in [11] *clock-reversal transformation*. The clock-reversed solution will be a solution of the Hamilton equations corresponding to the other sheet of the Hamiltonian constraint. This operation does correspond now to the operation of "playing the film backwards". In the extended configuration space which includes the physical clock, the graph of the solution obtained by means of this operation coincides with the original one, being the only difference the sense of evolution. In other words, it is now the same film (i.e., the same curve in the extended configuration space) played in the reversed sense.

Given a solution of the equations of motion in one of the sheets of the Hamiltonian constraint, it is then possible to construct three others related solutions: the motion-reversal (in the same sheet as the original one), the clock-reversal (in the other sheet), and the motion reversal of the clock-reversed solution (in the other sheet). Summarizing, we have two solutions in each sheet.

The main difference between the linear formalism for parametrized systems and the quadratic one is that the latter is invariant under clock-reversal transformations. In general, any dynamical trajectory (which carries then a *fixed* sense of evolution) can be considered as a static one by considering the clock in an extended configuration space. Inversely, any static trajectory (without a *fixed* sense of evolution) can be "temporalized" by considering a monotonously increasing variable along the trajectory as a physical clock. We could say that if the linear formalism is the proper formalism for considering the solutions of a dynamical system as static trajectories in an extended configuration space, the quadratic formalism is the proper one for expressing the static solutions as dynamical ones. In the first case the dynamical solution that one wants to render static has its own sense of evolution. It is then unnecessary to consider a Hamiltonian constraint with a two sheet structure in the process of parametrization. On the contrary, in the second case there is not a preferred sense of evolution for the "temporalization" of the static trajectory. Given that all the canonical variables are true degrees of freedom, there is not a "true" time disguised among them. This absence of a true time means that there is not a preferred sense of evolution for unfolding the trajectory. The Hamiltonian constraint must include then the two possible senses of evolution, which means that it must be quadratic in all its momenta.

In a quantum mechanical context these two operations assume very simple forms. Given a particular solution $\Psi_+(q^\mu, q^0)$ to the Schrödinger equation

$$i\frac{\partial}{\partial q^0}\Psi_+\left(q^\mu, q^0\right) = \hat{h}\Psi_+\left(q^\mu, q^0\right) \tag{68}$$

corresponding to the sheet $p_0 + h = 0$ ($t = q^0$), its motion-reversed solution is given by

$$\Psi_+^{mr}\left(q^\mu, q^0\right) = T\Psi_+\left(q^\mu, -q^0\right) \tag{69}$$

where T is an antiunitary operator which, in coordinate representation, is equal to the complex conjugation operator $T\Psi(q) = \Psi^*(q)$. We could say that if the T operation changes the sense of all variables $q^i = \{q^\mu, q^0\}$, the substitution $q^0 \to -q^0$ cancels this change *only* for the variable q^0. It is important to remark that this motion reversed solution Ψ_+^{mr} is a solution of the same Schrödinger equation (68). On the contrary, the quantum version of the clock-reversal transformation is performed *only* by the action of the antiunitary operator T (now we want to change *also* the sense of q^0):

$$\Psi_-^{cr}\left(q^\mu, q^0\right) = T\Psi_+\left(q^\mu, q^0\right). \tag{70}$$

This clock reversed solution Ψ_-^{cr} is not a solution of the Schrödinger equation (68) but a solution of the Schrödinger equation corresponding to the other sheet of the Hamiltonian constraint ($p_0 - h = 0$, $t = -q^0$):

$$-i\frac{\partial}{\partial q^0}\Psi_-\left(q^\mu, q^0\right) = \hat{h}\Psi_-\left(q^\mu, q^0\right). \tag{71}$$

It is important to remark, contrary to what is usually believed, that each sheet of the Hamiltonian constraint ($p + h = 0$ and $p - h = 0$) corresponds to positive energies solutions. What changes from one sheet to the other one is the sign of time, not the sign of the energy. Each sheet corresponds to each possible choice in the direction in which the static trajectory can be unfolded, being in both cases positive energy solutions. In fact, the stationary solutions to the Schrödinger equation (68) are

$$\Psi_+^E\left(q^\mu, q^0\right) = e^{-iEq^0}\varphi(q_\mu) = e^{-iEt}\varphi(q_\mu), \tag{72}$$

while the stationary solutions to the Schrödinger equation (71) are

$$\Psi_-^E\left(q^\mu, q^0\right) = e^{iEq^0}\varphi^*(q_\mu) = e^{-iEt}\varphi^*(q_\mu), \tag{73}$$

which shows that both sets of solutions are positive energy solutions for the two possible choices of the physical clock ($t = q^0$ or $t = -q^0$).

Following this new interpretative framework, the boundary conditions for the space of solutions of the Wheeler–DeWitt equation can be interpreted as the *symmetry breaking* of the clock-reversal invariance of a theory described by a Hamiltonian constraint with a two sheet structure. The boundary conditions proposed in Section 3.4 separate the space of solutions in those quantum states going forward in the time $t = q^0$ from those quantum states going forward in the time $t = -q^0$, being these two subspaces related by the clock-reversal operator T. The fact that the Wheeler-DeWitt operator H is a real operator has as a consequence that, given a certain solution Ψ, the associated function Ψ^* will be also a solution, being these solutions linearly independent. The space of solutions S of the Wheeler–DeWitt equation can then be decomposed as the direct sum $S = C \oplus C^*$. In the case of the solution (37) to the Wheeler–DeWitt equation (36), the factor depending on ϕ could be expressed equivalently as a linear combination of the modified Bessel functions K_ν and I_ν or as a linear combination of the modified Bessel functions I_ν and $I_{-\nu}$. But while the functions K_ν and I_ν are not complex conjugated of each other, the functions $I_{\pm\nu}$ do satisfy the property $I_\nu = I^*_{-\nu}$. As a result, the decompositions $S = K_\nu \oplus I_\nu$ and

$S = I_\nu \oplus I_{-\nu}$ are not equivalent, being the latter the correct decomposition for imposing the proposed boundary conditions. If one can find a decomposition of the space of solutions of the form $S = C \oplus C^*$, then the problem of imposing boundary conditions on this space of solutions is solved.

An important difference between the clock-reversal and the motion-reversal transformations is that a theory with a time dependent reduced Hamiltonian $h(t)$, even if it is not symmetric under a motion-reversal transformation, it is still symmetric under a clock-reversal transformation. This means that even if it is not possible to reduce the system by means of a time independent reduced Hamiltonian h, the space of solutions of the Wheeler-DeWitt equation must continue to be a direct sum of two subspaces related by the antiunitary operator T.

5 Conclusion

The quantization of string cosmologies is of particular interest because they allow for new points of view about the earliest stages of the universe [24, 23]. In the present work we have focused our attention in the formal aspects of the problem within the minisuperspace approximation. We shall now enumerate the central points of the whole discussion. As it is well known from quantum cosmology, the most important fact for the comprehension of the canonical structure and quantization of dilaton cosmological models is the presence of the so called *Hamiltonian constraint*. As it was explained in these notes, a theory with this kind of constraint lacks of a definition of a "true" time variable. A theory with such a canonical structure can not be directly interpreted in terms of an evolution in time of its degrees of freedom. In order to "temporalize" the theory it is necessary to select a *physical clock*, i.e., a monotonously increasing variable which could serve as a time parameter for measuring the evolution of the rest of the canonical variables. The notion of a physical clock was not interpreted in terms of a recovered "real" hidden time, but in terms of a particular degree of freedom which satisfies certain properties which let us use it as a globally well defined clock (it is then not necessary to search for a physical clock *only* among the geometrical degrees of freedom). We have analyzed the consequences of this interpretation on the so called *parametrized formalism*, which is usually used for understanding the canonical structure of a theory with a Hamiltonian constraint *linear* in one of its momenta. It was then analyzed a modification of this parametrized formalism in order to better describe the *quadratic* Hamiltonian constraint of the gravitational models that we are studying.

If a physical clock can be separated, then the theory can be quantized in a straightforward manner by means of the parabolic Schrödinger equations associated to each sheet of the Hamiltonian constraint. If one does not know how to separate a physical clock, one can still quantize the theory by following the so called Dirac method for quantizing constrained theories. Following this method, the physical quantum states have to satisfy the operator version of the quadratic Hamiltonian constraint (Wheeler–DeWitt equation). The problem of this method is that, lacking of a definition of time, we do not know neither how to interpret the corresponding solutions in terms of a conserved positive-definite probabilities, nor how to impose boundary conditions on the space of solutions of the Wheeler–DeWitt equation. In these notes we have studied certain models which can be quantized by following both methods (Schrödinger and Wheeler–DeWitt quantizations). The important notion of a *quantum canonically transformation* was introduced in order to relate the solutions obtained by both methods. This models are of great utility in order to propose general boundary conditions for those cases for which we do not know how to reduce the model by selecting a physical clock.

Finally we have discussed the temporal symmetries which are characteristic of the theories with a quadratic Hamiltonian constraint. Besides the *motion reversal transformation* which appears naturally in ordinary (non constrained) classical and quantum systems, we have described the so called *clock reversal transformation*. This symmetry arises as a consequence of the existence of a Hamiltonian constraint with a two sheet structure, being this symmetry associated to the passage from one sheet to the other one. We have conjectured that this symmetry could play an important role for the clarification of the boundary conditions to be imposed on the space of solutions of the Wheeler–DeWitt equation.

6 Appendix A

The condition for a function $t(q^i, p_i)$ to be a global phase time (that its Poisson bracket with the Hamiltonian constraint is positive definite) can be understood as follows. If we define the Hamiltonian vector field

$$\mathrm{H}^A = (\mathrm{H}^q, \mathrm{H}^p) = \left(\frac{\partial \mathcal{H}}{\partial p}, -\frac{\partial \mathcal{H}}{\partial q}\right), \tag{74}$$

then the condition

$$[t, \mathcal{H}] > 0$$

is equivalent to

$$\mathrm{H}^A \frac{\partial t}{\partial x^A} > 0,$$

with $x^A = (q^i, p_i)$. This means that $t(q^i, p_i)$ monotonically increases along any dynamical trajectory. Each surface $t = constant$ in the phase space is then crossed only once by any dynamical trajectory (so that the field lines of the Hamiltonian vector field are open). The successive states of the system can then be parametrized by $t(q^i, p_i)$.

If we explicitly write the constraint, the condition for the existence of a global time which depends only on the coordinates (*intrinsic time*) reads

$$[t(q^i), \mathcal{H}] = [t(q^i), G^{ik} p_i p_k] = 2\frac{\partial t}{\partial q^i} G^{ik} p_k > 0. \tag{75}$$

If the supermetric has a diagonal form and one of the momenta vanishes at a given point of phase space, then no function depending only on its conjugated coordinate can be a global phase time. For a constraint whose potential can be zero for finite values of the coordinates, the momenta p_k can be all equal to zero at a given point, and $[t(q^i), \mathcal{H}]$ can vanish. Hence an *intrinsic time* $t(q^i)$ can be identified only if the potential in the constraint has a definite sign. In the most general case a global phase time should be a function which depends also on the canonical momenta; in this case it is said that the system has an *extrinsic time* $t(q^i, p_i)$, because the momenta are related to the extrinsic curvature (see, for example [6]).

It must be noted that scaling a given constraint by a positive function does not affect the definition of time, that is, if $[t, \mathcal{H}] > 0$, then for $H = F(q)\mathcal{H}$, $F(q) > 0$ we also have $[t, H] > 0$ (we have used this property many times in these notes). From the point of view of quantization, however, things are less trivial, and the question arises about the validity of scaling the Hamiltonian. In this sense, it is important to recall that it can be shown that an operator ordering exists such that both constraints H and $\mathcal{H}$ are equivalent at the quantum level. Let us consider a generic constraint

$$\mathcal{H} = e^{bq_1}\left(-p_1^2 + p_2^2 + \zeta e^{aq_1+cq_2}\right) = 0,$$

which contains an ambiguity associated to the fact that the most general form of the first term should be written

$$-e^{Aq_1} p_1 e^{(b-A-C)q_1} p_1 e^{Cq_1}$$

(so that A and C parametrize all possible operator orderings). It is simple to verify that the constraint with the most general ordering differs from that with the trivial ordering in two terms, one linear and one quadratic in $\hbar$, and that these terms vanish with the choice $C = b - A = 0$. Therefore, the Wheeler–DeWitt equation resulting from the scaled constraint $H = e^{-bq_1}\mathcal{H} = 0$ is right in the sense that it corresponds to a possible ordering of the original constraint $\mathcal{H}$.

7 Appendix B

Start from the parametrized action

$$S[q^i, p_i, N] = \int_{\tau_1}^{\tau_2} \left(p_i \frac{dq^i}{d\tau} - N\mathcal{H}\right) d\tau \tag{76}$$

where $\mathcal{H} = 0$ on the constraint surface. Consider the τ-independent Hamilton–Jacobi equation

$$G^{ij} \frac{\partial W}{\partial q^i} \frac{\partial W}{\partial q^j} + V(q) = E \tag{77}$$

which results by substituting $p_i = \partial W / \partial q^i$ in the Hamiltonian. A complete solution $W(q^i, \alpha_\mu, E)$ (see for example [36]) obtained by matching the integration constants (α_μ, E) to the new momenta $(\overline{P}_\mu, \overline{P}_0)$ generates a canonical transformation

$$p_i = \frac{\partial W}{\partial q^i}, \quad \overline{Q}^i = \frac{\partial W}{\partial \overline{P}_i}, \quad K = N\overline{P}_0 = N\mathcal{H} \tag{78}$$

which identifies the constraint $\mathcal{H}$ with the new momentum $\overline{P}_0$. The variables $(\overline{Q}^\mu, \overline{P}_\mu)$ are conserved observables because $[\overline{Q}^\mu, \mathcal{H}] = [\overline{P}_\mu, \mathcal{H}] = 0$, so that they are not suitable for characterizing the dynamical evolution. A second transformation generated by the function

$$F = P_0 \overline{Q}^0 + f(\overline{Q}^\mu, P_\mu, \tau) \tag{79}$$

gives

$$\overline{P}_0 = P_0 \quad \overline{P}_\mu = \frac{\partial f}{\partial \overline{Q}^\mu} \quad Q^0 = \overline{Q}^0 \quad Q^\mu = \frac{\partial f}{\partial P_\mu} \tag{80}$$

and a new non vanishing Hamiltonian $K = NP_0 + \partial f/\partial \tau$, so that (Q^μ, P_μ) are non conserved observables because $[Q^\mu, \mathcal{H}] = [P_\mu, \mathcal{H}] = 0$ but $[Q^\mu, K] \neq 0$ and $[P_\mu, K] \neq 0$; we have, instead, that $[Q^0, \mathcal{H}] = [Q^0, P_0] = 1$, and then Q^0 can be used to fix the gauge. The transformation $(q^i, p_i) \to (Q^i, P_i)$ leads to the action

$$\mathcal{S}[Q^i, P_i, N] = \int_{\tau_1}^{\tau_2} \left(P_i \frac{dQ^i}{d\tau} - NP_0 - \frac{\partial f}{\partial \tau} \right) d\tau \tag{81}$$

which contains a linear and homogeneous constraint $P_0 = 0$ and a non zero (true) Hamiltonian $\partial f/\partial \tau$. Then we have obtained the action of an ordinary gauge system. Fixing the gauge on this system defines a particular foliation of space-time for the associated cosmological model originally described by the parametrized action.

In terms of the original variables the action $\mathcal{S}$ reads

$$S[q^i, p_i, N] = \mathcal{S}[q^i, p_i, N] + \left[\overline{Q^i} \overline{P_i} - W + Q^\mu P_\mu - f \right]_{\tau_1}^{\tau_2} \tag{82}$$

We see that $\mathcal{S}$ and S differ only in end point terms; thus both actions yield the same dynamics.

References

[1] Antoniadis I., Bachas C., Ellis J. and Nanopoulos D. V., *Phys. Lett.* **B221**, 393, (1988).

[2] Baleanu D. and Guler Y., *J. Phys.* **A34**, 73 (2001).

[3] Baleanu D., *Reparamerization Invariance and HamiltonñJacobi formalism* (2004), hep-th/0412050.

[4] Barbour, J. B., *Class. Quant. Grav.* **11**, 2853 (1994); *Class. Quant. Grav.* **11**, 2875 (1994).

[5] Barbour J. B., *The end of time*, Oxford University Press (1999)

[6] Barvinsky A. O., *Phys. Rep.* **230**, 237 (1993).

[7] Beluardi S. C. and Ferraro R., *Phys. Rev.* **D52**, 1963 (1995).

[8] Blyth W. F. and Isham C. J., *Phys. Rev.* **D11**, 768 (1975).

[9] Brown J. D. and Kuchar K. V., *Phys. Rev.* **D51**, 5600 (1995).

[10] Carlip S., *Class. Quant. Grav.* **11**, 31 (1994).

[11] Castagnino M., Catren G. and Ferraro R., *Class. Quant. Grav.* **19**, 4729 (2002).

[12] Catren G. and Ferraro R., *Phys. Rev.* **D63**, 023502 (2001).

[13] Cavaglia M., *Int. J. Mod. Phys.* **D8**, 101 (1999).

[14] Cavaglia M. and De Alfaro A., *Gen. Rel. Grav.* **29**, 773 (1997).

[15] Cavaglia M. and Ungarelli C., *Class. Quant. Grav.* **16**, 1401 (1999).

[16] Cavaglia M. and Ungarelli C., *Nucl. Phys. Proc. Suppl.* **88**, 355 (2000).

[17] De Cicco H. and Simeone C., *Gen. Rel. Grav.* **31**, 1225 (1999).

[18] De Cicco H. and Simeone C., *Int. J. Mod. Phys.* **A14**, 5105 (1999).

[19] DeWitt B. S., *Phys. Rev.* **160**, 1113 (1967).

[20] Dirac P. A. M., *Lectures on Quantum Mechanics,* Belfer Graduate School of Science, Yeshiva University, New York (1964).

[21] Ferraro R. and Simeone C., *J. Math. Phys.* **38**, 599 (1997).

[22] Ferraro R., *Grav. Cosm.* **5**, 195 (1999).

[23] Gasperini M., *Class. Quant. Grav.* **17** R1 (2000). Gasperini M., in *Proceedings of the 2nd SIGRAV School on Gravitational Waves in Astrophysics, Cosmology and String Theory, Villa Olmo, Como*, edited by V. Gorini, hep-th/9907067.

[24] Gasperini M. and Veneziano G., *Phys. Rept.* **373**, 1 (2003).

[25] Ghandour G. I., *Phys. Rev.* **D35**, 1289 (1987).

[26] Giribet G. and Simeone C., *Mod. Phys. Lett.* **A16**, 19 (2001).

[27] Goldwirth D. S., and Perry M. J., *Phys. Rev.* **D49**, 5019 (1994).

[28] Hájíček P., *Phys. Rev.* **D34**, 1040 (1986).

[29] Halliwell J. J., in *Introductory Lectures on Quantum Cosmology*, Proceedings of the Jerusalem Winter School on Quantum Cosmology and Baby Universes, edited by T. Piran, World Scientific, Singapore (1990).

[30] Henneaux M., Teitelboim C. and Vergara J. D., *Nucl. Phys.* **B387**, 391 (1992).

[31] Henneaux M. and Teitelboim C., *Quantization of Gauge Systems,* Princeton University Press, New Jersey (1992).

[32] Kuchar K. V., *Phys. Rev.* **D4**, 955 (1971).

[33] Kuchar K. V., in *Quantum Gravity 2: A Second Oxford Symposyum*, edited by C. J. Isham, R. Penrose and D. W. Sciama, Clarendon Press (1981).

[34] Kuchar K. V., in *Proceedings of the 4th Canadian Conference on General Relativity and Relativistic Astrophysics*, edited by G. Kunstatter, D. Vincent and J. Williams, World Scientific, Singapore (1992).

[35] Kuchar K. V., in *General Relativity and Gravitation 1992*, Proceedings of the 13th International Conference on General Relativity and Gravitation, Córdoba, Argentina, edited by R. Gleiser, C. N. Kozameh and O. M. Moreschi, IOP Publishing, Bristol (1993).

[36] Landau L. D. and Lifshitz E. M., *Mechanics*, Pergamon Press, Oxford (1960).

[37] Moncrief V. and Ryan M. P., *Phys. Rev.* **D44**, 2375 (1991).

[38] Polchinski J., *String Theory I, An Introduction to the Bosonic String*, Cambridge University Press, Cambridge (1998).

[39] Polyakov A. M., *Phys. Lett.* **B103**, 207 (1981).

[40] Savchenko V. A., Shestakova T. P. and Vereshkov G. M., *Grav. Cosm.* **7**, 18 (2001), gr-qc/9809086.

[41] Savchenko V. A., Shestakova T. P. and Vereshkov G. M., *Grav. Cosm.* **7**, 102 (2001), gr-qc/9810035.

[42] Shestakova T. P. and Simeone C., *Grav. Cosm.* **10**, 161 (2004).

[43] Simeone C., *J. Math. Phys.* **39**, 3131 (1998).

[44] Simeone C., *J. Math. Phys.* **40**, 4527 (1999).

[45] Simeone C., *Deparametrization and Path Integral Quantization of Cosmological Models*, World Scientific Lecture Notes in Physics 69, World Scientific, Singapore (2002).

[46] Simeone C., *Phys. Lett.* **A310**, 143 (2003).

[47] Tkach V. I., Pashnev A. and Rosales J. J., *Reparametrization Invariance and the Schrödinger equation* Preprint JINR E2-99-311 (1999), hep-th/9912282.

[48] Tseytlin A. A., *Class. Quant. Grav.* **9**, 979, (1992).

[49] Tseytlin A. A. and Vafa C., *Nucl. Phys.* **B372**, 443, (1992).

[50] Varadarajan M., *Phys. Rev.* **D70**, 084013 (2004).

[51] Veneziano G., *Phys. Lett.* **B265**, 387 (1991).

[52] Veneziano G., *String Cosmology: The pre-big bang scenario*, Lectures delivered in Les Houches (1999), hep-th/0002094.

[53] York J. W., *Phys. Rev. Lett.* **28**, 1082 (1972).

In: Trends in General Relativity and Quantum Cosmology
Editor: Charles V. Benton, pp. 79-101
ISBN 1-59454-794-7
© 2006 Nova Science Publishers, Inc.

Chapter 5

QUANTUM MECHANICS EMERGING FROM "TIMELESS" CLASSICAL DYNAMICS

Hans-Thomas Elze
Dipartimento di Fisica, Università di Pisa,
Largo Pontecorvo 3, I–56127 Pisa, Italia
E-mail address: elze@df.unipi.it

Abstract

We study classical Hamiltonian systems in which the intrinsic proper time evolution parameter is related through a probability distribution to the physical time, which is assumed to be discrete. In this way, a physical clock with discrete states is introduced, which presently is still treated as decoupled from the system. This is motivated by the recent discussion of "timeless" reparametrization invariant models, where discrete physical time has been constructed based on quasi-local observables. Employing the path-integral formulation of classical mechanics developed by Gozzi et al., we show that these deterministic classical systems can be naturally described as unitary quantum mechanical models. We derive the emergent quantum Hamiltonian in terms of the underlying classical one. Such Hamiltonians typically need a regularization – here performed by discretization – in order to arrive at models with a stable groundstate in the continuum limit. This is demonstrated in several examples, recovering and generalizing a model advanced by 't Hooft.

PACS: 03.65.Ta, 04.20.-q, 05.20.-y

1 Introduction

Since its very beginnings, there has been a long series of speculations on the possibility of deriving quantum theory from more fundamental dynamical structures, possibly deterministic ones. Famous is the discussion by Einstein, Podolsky and Rosen. This led to the EPR paradox, which in turn was interpreted by its authors as indicating the need for a more complete fundamental theory [1]. However, just as numerous have been attempts to prove no-go theorems prohibiting exactly such "fundamentalism". This culminated in the studies of Bell, leading to the Bell inequalities [2]. The paradox as well as the inequalities have come under experimental scrutiny in recent years, confirming the predictions of quantum mechanics in laboratory experiments on scales very large compared to the Planck scale.

However, to this day, the feasible experiments cannot rule out the possibility that quantum mechanics emerges as an effective theory only on sufficiently large scales and can indeed be based on more fundamental models.

Motivated by the unreconciled clash between general relativity and quantum theory, 't Hooft has strongly argued in favour of model building in this context [3] (see also further references therein). He has shown in individual examples the emergence of the usual Hilbert space structure and unitary evolution in deterministic classical models in an appropriate large-scale limit. Particular emphasis has been placed on the observation that while it is relatively easy to arrive at a Hilbert space formulation for classical dynamics, it is difficult to obtain emergent Hamiltonians with a spectrum bounded from below, i.e., having a well-defined groundstate.

Various further arguments for deterministically induced quantum features have recently been proposed, for example, in Refs. [4, 5, 6, 7], in the context of statistical systems, of considerations related to quantum gravity, and of matrix models, respectively.

Our aim here is to contribute to this line of research by reporting a rather large class of classical deterministic systems which yield to a proper quantum mechanical description. This is based on our recent work on time-reparametrization invariant models [8, 9]. In particular we have introduced the construction of a discrete physical time for such "timeless" systems, which will be the starting point of the present developments. In the following we briefly summarize essential aspects.

1.1 Discrete physical time in "timeless" classical models

Similarly to common gauge theories, such as those of the Standard Model of particle physics, reparametrization invariant systems show invariance under a kind of gauge transformations. In the most general case of diffeomorphism invariant theories, such as general relativity or string theory, this amounts to invariance under general coordinate transformations.

We limit ourselves to time-reparametrization invariance here, which for our purposes can be expressed as invariance of the dynamics under arbitrary transformations:[1]

$$t \longrightarrow t', \text{ with } t \equiv f(t') . \tag{1}$$

Details of the corresponding constrained Lagrangian dynamics can be found in Refs. [8, 9] for the respective models.

Similarly as Gauss' law in electrodynamics, for example, a most important consequence of time-reparametrization invariance is a (weak) constraint, which states that the Hamiltonian has to vanish (on the solutions of the equations of motion).[2] Since the Hamiltonian commonly is the generator of time evolution of the system, this is what has led to name this type of systems "timeless". A Newtonian external time parameter does not exist. This

---

[1] Some restrictions are imposed in Refs. [8, 9] on physical grounds, such as differentiability and monotonicity of the function f.

[2] In the case of the free relativistic particle the invariance amounts to invariance under reparametrization of the proper time parameter and the ensuing constraint is the familiar mass-shell constraint on its momentum.

becomes problematic when trying to quantize such systems, since a standard Schrödinger equation does not exist either. In particular, the Wheeler-DeWitt equation, $\hat{H}|\psi\rangle = 0$, epitomizes the intrinsic problems of quantum gravity, seen from this perspective.

Numerous approaches have been tried to resolve this (in)famous "problem of time". For our purposes, it may suffice to mention one, in order to contrast it with our proposal.

In Refs. [10, 11, 12] it has been assumed that *global* features of a suitably parametrized trajectory of the system, are accessible to the observer. This makes it possible, in principle, to express the evolution of an arbitrarily selected degree of freedom relationally in terms of others. Naively speaking, the question "What time is it?" is replaced by "What is the value x of observable X, *when* observable Y has value y?". Thereby the Hamiltonian and possibly additional constraints have been eliminated in favour of Rovelli's "evolving constants of motion".

In distinction, we insist on a *local* description. We have shown that for a particle with time-reparametrization invariant dynamics, be it relativistic or nonrelativistic, one can define quasi-local observables which characterize the evolution in a gauge invariant way [8, 9].

Essentially, we employ some of the degrees of freedom of the system to trigger a localized "detector". This can be defined in an invariant way. It amounts to attributing to an observer the capability to count discrete events. – In passing we remark that a similar approach, avoiding the notion of time altogether and replacing it by counts of idealized coincidence detectors in phase space has recently also been put forth by Halliwell and Thorwart [13]. – In any case, the detector counts present an observable measure of time. This *physical time is discrete*.

This result can be understood in a different way, by noting that our construction is practically based on a Poincaré section, or subsection thereof, which reflects an ergodic if not periodic aspect of the dynamics – quite analogous to a pendulum which triggers a coincidence counter each "time" it passes through its equilibrium position. Reparametrization invariance strongly limits the information which can be extracted from it with respect to a complete trajectory. This is the underlying reason that a physical time based on local observations (clock readings) necessarily is discrete.[3]

Independently of our physical motivation of discrete time, we remark that the possibility of a fundamentally discrete time (and possibly other discrete coordinates) has been explored before, ranging from an early realization of Lorentz symmetry in such a case [14] to detailed explorations of its consequences and consistency in classical mechanics, quantum field theory, and general relativity [15, 16, 17]. Further recent developments are discussed in the review of Ref. [18]; particularly in relation to fundamental or induced violations of Lorentz symmetry, which are believed to come within experimental reach in the near future.

So far, however, no classical physical models giving rise to such discreteness were proposed. *Quantization as an additional step* – which results in discreteness of coordinates in various cases – has always been performed as usual.

---

[3]This does not conflict with certain cosmological models where a scalar matter field coupled to gravity is invoked as a continuous time variable. Here, the need to know the scalar field globally can be circumvented by assuming its homogeneity.

1.2 Does discrete time induce quantum mechanical features?

It is the purpose of the present study to reach a qualified 'Yes', answering this question. We hope this will contribute to the study of potentially deterministic substructures leading to quantum mechanics as an emergent theory.

Based on the findings of Refs. [8, 9], respectively, we have argued there that those discrete-time models can be mapped on a cellular automaton studied by 't Hooft before [3]. With the help of the algebra of $SU(2)$ generators, it has been shown that these models actually reproduce the quantum mechanical harmonic oscillator in a suitably defined large-scale limit.

We will come back to variants of the earlier results in Section 6. However, in the following main parts of this chapter, we attempt to show more generally that due to inaccessability of globally complete information on trajectories of the system, the evolution of remaining degrees of freedom appears as in a quantum mechanical model when described in relation to the discrete physical time.

We may call this "stroboscopic" quantization: when a continous physical time is not available but a discrete one is – like reading an analog clock under a stroboscopic light – then states of the system which fall in between subsequent clock "ticks" cannot be resolved. (Of course, evolution in the unphysical parameter time is continous in the constrained Lagrangian models we refer to.) Such unresolved states form equivalence classes which can be identified with primordial Hilbert space states [3, 4, 8, 9]. The residual dynamics then has to describe the evolution of these states through discrete steps. Under favourable circumstances, this results in unitary quantum mechanical evolution, as we shall see.

Our present aim is to show that this occurs quite generally in classical Hamiltonian systems, if time is discrete. We will presently simplify the situation by assuming that the physical time can be related by a probability distribution to the proper time of the equations of motion. Explicit examples of such behaviour can be found in Refs. [8, 9], when the clock degrees of freedom evolve independently of the rest of the system, apart from the Hamiltonian constraint.

Thus, while the investigation of the coupled system-clock dynamics is presently under study [19], here we make the approximation that corresponding "backreaction" effects are small. This feature, besides characterizing a good clock, may serve as an appropriate simplification in our exploratory steps. In the concluding section, we will briefly comment about extensions, where the prescribed probabilistic mapping of physical onto proper time shall be abandoned in favour of a selfconsistent treatment. A closed system, of course, has to include its own clock, if it is not entirely static, reflecting the experience of an observer in the universe.

Finally, in order to put our approach into perspective, we remark that there is clearly no need to follow such construction leading to a discrete physical time in ordinary mechanical systems or field theories, where time is an external classical parameter. However, assuming for the time being that truly fundamental theories will turn out to be diffeomorphism invariant, adding further the requirement of the observables to be quasi-local,[4] when describing the evolution, then such an approach seems natural, which promises to lead to quantum

---
[4]For example, a closed string loop representing a fundamental length might define the ultimate resolution of distance measurements and, thus, condition the notion of locality.

mechanics as an emergent description or effective theory on the way.

This chapter is organized as follows. Section 2 provides a selfcontained brief summary of the path-integral formulation of classical mechanics. We will employ this as a convenient tool to formulate our approach. In later developments one might also introduce a cooresponding operator formalism, for example, considering as a starting point the early work of Koopman and von Neumann [20].

In Section 3 the relevant to-be-quantum states are introduced, i.e. equivalence classes of states of the underlying classical system, as we mentioned. Their evolution is studied in Section 4, leading to an emergent Hamilton operator and associated Schrödinger equation under circumstances to be discussed there. In Section 5 the calculation of physical time dependent observables, or expectation values of corresponding operators, is related to the states.

Section 6 presents some simple examples of the emergent Hamiltonians and the calculation of their spectra. We show that – under the present simplifying assumptions – these operators need to be regularized, in order to represent acceptable quantum models, with a stable groundstate in particular. Here we achieve this by discretization. We speculate that its apparent arbitrariness might be removed by a future selfconsistent treatment of clock-system interactions, which will lead to dissipative effects on small scales and could define a unique quantum system for each classical model with discrete time. The concluding Section 7 presents a brief summary of the present chapter and points out several open problems.

2 Classical mechanics via path-integrals

Classical mechanics can be cast into path-integral form, as originally developed by Gozzi, Reuter and Thacker [21], and with recent addenda reported in Ref. [22]. While the original motivation has been to provide a better understanding of geometrical aspects of quantization, we presently use it as a convenient tool. We refer the interested reader to the cited references for details, on the originally resulting extended (BRST type) symmetry in particular. Here we suitably incorporate time-reparametrization invariance, assuming equations of motion written in terms of proper time (as in our earlier examples [8, 9]).

Let us begin with a $(2n)$-dimensional classical phase space $\mathcal{M}$ with coordinates denoted collectively by $\varphi^a \equiv (q^1, \ldots, q^n; p^1, \ldots, p^n)$, $a = 1, \ldots, 2n$, where q, p stand for the usual coordinates and conjugate momenta. Given the proper-time independent Hamiltonian $H(\varphi)$, the equations of motion are:

$$\frac{\partial}{\partial \tau}\varphi^a = \omega^{ab}\frac{\partial}{\partial \varphi^b}H(\varphi) , \qquad (2)$$

where ω^{ab} is the standard symplectic matrix and τ denotes the proper time; summation over indices appearing twice is understood.

To the equation of motion we add the (weak) Hamiltonian constraint, $C_H \equiv H(\varphi) - \epsilon \simeq 0$, with ϵ a suitably chosen parameter. This constraint has to be satisfied by the solutions of the equations of motion. Generally, it arises in reparametrization invariant models, similarly as the mass-shell constraint in the case of the relativistic particle [9]. It is necessary when the Lagrangian time parameter is replaced by the proper time in the equations of motion.

In this way, an arbitrary "lapse function" is eliminated, which otherwise acts as a Lagrange multiplier for this constraint.

We remark that field theories can be treated analogously, considering indices a, b, etc. as continuous variables.

Starting point for our following considerations is the *classical* generating functional,

$$Z[J] \equiv \int_H \mathcal{D}\varphi \; \delta[\varphi^a(\tau) - \varphi^a_{cl}(\tau)] \exp(i \int d\tau \; J_a \varphi^a) \; , \tag{3}$$

where $J \equiv \{J_{a=1,...,2n}\}$ is an arbitrary external source, $\delta[\cdot]$ denotes a Dirac δ-functional, and φ_{cl} stands for a solution of the classical equations of motion satisfying the Hamiltonian constraint; its presence is indicated by the subscript "H" on the functional integral. It is important to realize that $Z[0]$, as it stands, gives weight 1 to any classical path satisfying the constraint and zero otherwise, integrating over all initial conditions.

Using the functional equivalent of $\delta(f(x)) = |df/dx|_{x_0}^{-1} \cdot \delta(x - x_0)$, the δ-functional under the integral for Z can be replaced according to:

$$\delta[\varphi^a(\tau) - \varphi^a_{cl}(\tau)] \Longrightarrow \delta[\partial_\tau \varphi^a - \omega^{ab}\partial_b H] \det[\delta^a_b \partial_\tau - \omega^{ac}\partial_c \partial_b H] \; , \tag{4}$$

slightly simplifying the notation, e.g. $\partial_b \equiv \partial/\partial \varphi^b$. Here the modulus of the functional determinant has been dropped [21, 22].

Finally, the δ-functionals and determinant are exponentiated, using the functional Fourier representation and ghost variables, respectively. Thus, we obtain the generating functional in the convenient form:

$$Z[J] = \int_H \mathcal{D}\varphi \mathcal{D}\lambda \mathcal{D}c \mathcal{D}\bar{c} \; \exp\left(i \int d\tau (L + J_a \varphi^a)\right) \; , \tag{5}$$

which we abbreviate as $Z[J] = \int_H \mathcal{D}\Phi \; \exp(i \int d\tau L_J)$. The enlarged phase space is $(8n)$-dimensional, consisting of points described by the coordinates $(\varphi^a, \lambda_a, c^a, \bar{c}_a)$. The effective Lagrangian is now given by [21, 22]:

$$L \equiv \lambda_a \left(\partial_\tau \varphi^a - \omega^{ab}\partial_b H\right) + i\bar{c}_a \left(\delta^a_b \partial_\tau - \omega^{ac}\partial_c \partial_b H\right)c^b \; , \tag{6}$$

where $c^a, \bar{c}^a$ are anticommuting Grassmann variables. We remark that an entirely bosonic version of the path-integral exists [22].

This completes our brief review of how to put (reparametrization invariant) classical mechanics into path-integral form.

3 From discrete time to "states"

We recall from the motivation provided in Section 1.1 that discrete physical time t is constructed and based on the counting of suitably defined incidents. In particular, we have coincidences in mind, "when" points belonging to the trajectory of the system coincide with the position of an idealized detector. For concrete realizations of this procedure and study of its invariance as well as further intrinsic properties we refer the interested reader to Refs. [8, 9], and to Ref. [13] for a similar construction.

Thus, physical time is measured by a nonnegative integer multiple of some unit time, $t \equiv nT$, and is read off from a sort of localized track counting device.[5] For our present study details of the clock construction are of secondary importance. Rather we investigate the consequences of the discreteness of physical time. In particular, we want to demonstrate the existence of quantum mechanical features which may derive from it.

In any case, then, the proper time τ parametrizing the evolution should be calculable in terms of t. As can be observed, for example, in our earlier numerical simulations, this can be a formidable task, depending on the dynamics governing the underlying classical system [8, 9]. Therefore, an analytic approach for interacting clock-coupled-to-system models is beyond the scope of this chapter and considered elsewhere [19].

We will assume that the backreaction and memory effects which generally result from the coupling between clock and system are small and negligible for our present purposes. Then the relation between the discrete physical time t and the proper time τ of the equations of motion, Eq. (2), can be represented by a time independent normalized probability distribution P:

$$P(\tau;t) \equiv P(\tau - t) \equiv \exp\left(-S(\tau - t)\right) \ , \quad \int d\tau \, P(\tau;t) = 1 \ . \tag{7}$$

Note, in particular, that the perfect clock described in this way does not age with physical time. Eventually, we will also invoke the limiting case, $P(\tau;t) = \delta(\tau - t)$, i.e. a deterministic mapping between t and τ.

We remark that in the present situation, the Hamiltonian constraint needs to be applied to the system degrees of freedom only, while generally clock plus system are constrained as a whole. This is exemplified in all detail in our toy models [8, 9].

Correspondingly, we introduce the modified generating functional:

$$Z[J] \equiv \int_H d\tau_i d\tau_f \int \mathcal{D}\Phi \, \exp\left(i \int_{\tau_i}^{\tau_f} d\tau \, L_J - S(\tau_i - t_i) - S(\tau_f - t_f)\right) \ , \tag{8}$$

instead of Eq. (5), and using the condensed notation introduced there. In the present case, $Z[0]$ sums over all classical paths satisfying the constraint with weight $P(\tau_i;t_i) \cdot P(\tau_f;t_f)$, depending on their initial and final proper times, while all other paths get weight zero. In this way, the distributions of proper time values $\tau_{i,f}$ associated with the initial and final physical times, t_i and t_f, respectively, are incorporated.

Next, we insert $1 = \int d\tau P(\tau;t)$ into the expression for Z, with an arbitrarily chosen physical time $t_i < t < t_f$. This leads us to factorize the path-integral into two connected ones:

$$Z[J] = \int d\tau \, P(\tau;t) \cdot \int d\tau_f \int_H \mathcal{D}\Phi_> \, \exp\left(i \int_\tau^{\tau_f} d\tau' \, L_J^> - S(\tau_f - t_f)\right)$$

$$\cdot \int d\tau_i \int_H \mathcal{D}\Phi_< \, \exp\left(i \int_{\tau_i}^{\tau} d\tau'' \, L_J^< - S(\tau_i - t_i)\right)$$

$$\cdot \prod_a \delta(\varphi_>^a(\tau) - \varphi_<^a(\tau)) \ , \tag{9}$$

---

[5]In our earlier detailed examples we were motivated by the at present actively researched separation in higher dimensional (cosmological) models of bulk and brane degrees of freedom: interactions between both kinds of matter could lead to a more detailed picture of such countable incidents providing the basis of physical time.

where "<" and ">" refer to earlier and later than τ, respectively.

The ordinary δ-functions assure continuity of the classical paths in terms of the coordinates q^a, $a = 1, \ldots, n$ and momenta p^a, $a = 1, \ldots, n$, at the (distributed) proper time τ. This is necessary, in order to avoid a double-counting as compared to the original expression for Z, Eq. (8). Since any given classical path beginning at t_i and ending at t_f (with associated values $\tau_{i,f}$) is cut in two parts, without the continuity condition, the continuing part (">") would again have arbitrary initial conditions at the instant τ, unlike the corresponding original path where this is excluded by the Hamiltonian flow. To put it differently, without continuity, originally N independent paths would be factorized erroneously into N^2 independ ones.

Another remark is in order here. Due to the presence of the Hamiltonian constraints on both parts of the cut trajectory, one of the δ-functions is redundant. The resulting $\delta(0)$ may eventually be absorbed into the normalization of the states, which will be introduced now.

Exponentiating the δ-functions via Fourier transformation, the generating functional can indeed be interpreted as a scalar product of a state and its adjoint:

$$Z[J] = \int d\tau d\pi \ P(\tau) \langle t; t_f | \tau, \pi \rangle \langle \tau, \pi | t; t_i \rangle \ , \tag{10}$$

in "τ, π-representation"; here $d\pi \equiv \prod_a (d\pi_a/2\pi)$. The *state* is defined by the path-integral:

$$\langle \tau, \pi_a | t; t_0 \rangle \equiv \int d\tau' \int_H \mathcal{D}\Phi \ \exp\left(i \int_{\tau'}^{\tau+t} d\tau'' L_J - S(\tau' - t_0) + i\pi_a \varphi^a(\tau + t)\right) \ , \tag{11}$$

and, similarly, the *adjoint state*:

$$\langle t; t_0 | \tau, \pi_a \rangle \equiv \int d\tau' \int_H \mathcal{D}\Phi \ \exp\left(i \int_{\tau+t}^{\tau'} d\tau'' L_J - S(\tau' - t_0) - i\pi_a \varphi^a(\tau + t)\right) \ , \tag{12}$$

where t_0 represents an arbitrary physical time with which the respective paths begin, in the case of states, and end, in the case of adjoint states.

We observe a redundancy in designating the states, which depend on the sum of proper and physical time only. This arises by the shift of τ leading to Eq. (10), noting that the probability distribution P is not explicitly depending on the physical time, as we discussed.

Furthermore, there is a symmetry between states and adjoint states:

$$\langle t; t_0 | \tau, \pi \rangle = \langle \tau, \pi | t; t_0 \rangle^* \ , \tag{13}$$

if referred to the same reference time t_0, which is familiar in Hilbert space.

Finally, the scalar product of any two states can be defined by:

$$\langle t_2; t_f | t_1; t_i \rangle \equiv Z[J]^{-1} \int d\tau d\pi \ P(\tau) \langle t_2; t_f | \tau, \pi \rangle \langle \tau, \pi | t_1; t_i \rangle \ . \tag{14}$$

Of course, we have $\langle t; t_f | t; t_i \rangle = 1$, corresponding to Eq. (9), while for $t_2 \neq t_1$, the states generally are not orthogonal.

Closing this section, we point out that here the path-integrals have always been considered to include all classical paths. However, they can be restricted by imposing (final) initial conditions in the case of (adjoint) states. Since this has no effect on the following study of the time evolution of a generic state, there is presently no need to make explicitly use of this.

4 Unitary evolution

In order to learn about the time evolution of generic states (in the absence of a source, $J = 0$), we proceed similarly as in Eqs. (8)–(11).

Suitably inserting "1", as before, and splitting the path-integral (at time t) which defines a state (at time t'), we obtain:

$$\langle \tau', \pi' | t' \rangle = \int d\tau d\pi \ P(\tau) \langle \tau', \pi' | \widehat{U}(t', t) | \tau, \pi \rangle \langle \tau, \pi | t \rangle \ , \tag{15}$$

where henceforth we suppress to indicate the reference time t_0, see Eq. (11), in order to simplify the notation. Written as a matrix element of an evolution operator $\widehat{U}(t', t)$, the kernel which appears is:

$$\langle \tau', \pi' | \widehat{U}(t', t) | \tau, \pi \rangle \equiv \int_H \mathcal{D}\Phi \ \exp\left(i \int_{\tau+t}^{\tau'+t'} d\tau'' L + i\pi' \cdot \varphi(\tau'+t') - i\pi \cdot \varphi(\tau+t)\right) \ , \tag{16}$$

with the functional integral over *all paths* running between $\tau + t$ and $\tau' + t'$, subject to the Hamiltonian constraint; we abbreviate: $\pi \cdot \varphi \equiv \pi_a \varphi^a$.

Then, first of all, it is straightforward to establish the following composition rule:

$$\widehat{U}(t'', t') \cdot \widehat{U}(t', t) = \widehat{U}(t'', t) \ , \tag{17}$$

written here in the way of a matrix product which is to be interpreted as integration over intermediate variables, say $d\tau', d\pi'$, with appropriate weight factor $P(\tau')$. Then, the first integration, say over $d\tau'$, effectively removes a "1" – such as we inserted before, in order to factorize path-integrals. The second, say over $d\pi'$ reinstitutes δ-functions – such as those which first made their appearance in Eq. (9) – which serve to link the endpoint coordinates of one classical path to the initial of another. This leads to the result of the right-hand side.

Since the Hamiltonian constraint is a constant of motion, there is no need to constrain the path-integral representing the evolution operator. Integrating over intermediate variables removes all contributions violating the Hamiltonian constraint, provided we work with properly constrained states. This will be further discussed in the following section.

We observe in Eq. (16) that the physical-time dependence of the evolution operator (on t' and t) amounts to translations of proper time variables ($\tau' + t'$ and $\tau + t$) in the matrix elements. This simplicity, of course, is related to the analogous property of the states, see Eq. (11), which we discussed in the previous Section 3.

Tracing the steps backwards which led from Eq. (3) to Eq. (5) in Section 2, we can similarly rewrite here the functional integral of Eq. (16):

$$\langle \tau', \pi' | \widehat{U}(t', t) | \tau, \pi \rangle = \int \mathcal{D}\varphi \ \delta[\varphi^a(\tilde{\tau}) - \varphi_{cl}^a(\tilde{\tau})] \exp\left(i\pi' \cdot \varphi(\tau'+t') - i\pi \cdot \varphi(\tau+t)\right) \ , \tag{18}$$

where the paths parametrized by $\tilde{\tau}$ run between $\tau + t$ and $\tau' + t'$ and all initial conditions are integrated over.

Fixing momentarily the initial condition of the paths, $\varphi(\tau + t) = \varphi_i$, we select a particular path contributing in Eq. (18). By integrating over all initial conditions in the end,

$\int \mathrm{d}\varphi_i$, we recover the full expression. Again, this amounts to a suitable insertion of "1", a familiar procedure by now, which allows us to manipulate the above equation as follows:

$$\langle \tau', \pi' | \hat{U}(t', t) | \tau, \pi \rangle \tag{19}$$

$$= \int \mathcal{D}\varphi \int \mathrm{d}\varphi_i \, \delta(\varphi(\tau + t) - \varphi_i) \, \delta[\varphi^a(\tilde{\tau}) - \varphi^a_{cl}(\tilde{\tau})]$$

$$\times \exp\left(i\pi' \cdot \varphi(\tau' + t') - i\pi \cdot \varphi(\tau + t)\right)$$

$$= \int \mathrm{d}\varphi_i \, \exp\left(i\pi' \cdot \varphi_f - i\pi \cdot \varphi_i\right) \int \mathcal{D}\varphi \, \delta[\varphi^a(\tilde{\tau}) - \varphi^a_{cl}(\tilde{\tau})] ,$$

where φ_f denotes the value of $\varphi(\tilde{\tau})$ at the endpoint $\tau' + t'$ of the path singled out by the initial condition $\varphi(\tau + t) = \varphi_i$. This value is determined by the classical equations of motion and will be calculated shortly. Since the remaining functional integral equals one – only the one path with fixed initial conditon and weight 1 is contributing – we first obtain the intermediate result:

$$\langle \tau', \pi' | \hat{U}(t', t) | \tau, \pi \rangle = \int \mathrm{d}\varphi_i \, \exp\left(i\pi' \cdot \varphi_f - i\pi \cdot \varphi_i\right) , \tag{20}$$

where $\mathrm{d}\varphi_i \equiv \prod_{a=1}^{2n} \mathrm{d}\varphi_i^a$. Of course, the time dependent relation between φ_f and φ_i has to be enforced, in order to make this condensed expression explicit. This we do next.

We employ the Liouville operator,

$$\hat{\mathcal{L}} \equiv -\frac{\partial H}{\partial \varphi} \cdot \omega \cdot \frac{\partial}{\partial \varphi} , \tag{21}$$

where ω is the symplectic matrix, stemming from the Hamiltonian equations of motion (2). As is well known, the Liouville operator allows to generate a classical solution of these equations at any finite time, starting with a given initial condition. Applying this for our purposes yields:

$$\varphi_f \equiv \varphi(\tau' + t') = \exp[\hat{\mathcal{L}}(\tau' + t' - \tau - t)]\varphi(\tau + t) \equiv \exp[\hat{\mathcal{L}}(\tau' + t' - \tau - t)]\varphi_i . \tag{22}$$

Then, inserting into Eq. (20), we obtain the simple but central result:

$$\langle \tau', \pi' | \hat{U}(t', t) | \tau, \pi \rangle = \int \mathrm{d}\varphi \, \exp\left(i\pi' \cdot (\exp[\hat{\mathcal{L}}(\tau' + t' - \tau - t)]\varphi) - i\pi \cdot \varphi\right) \tag{23}$$

$$\equiv \mathcal{E}(\pi', \pi; \tau' + t' - \tau - t) , \tag{24}$$

where the by now superfluous subscript "i" has been omitted.

Using Eq. (23), one readily confirms Eq. (17) once again. In particular, then $\hat{U}(t|t') \cdot \hat{U}(t'|t) = \hat{U}(t|t)$, which is not diagonal, in general, in this τ, π-representation. We have: $U(\tau', \pi'; t | \tau, \pi; t) = \mathcal{E}(\pi', \pi; \tau' - \tau)$.

In order to proceed, we consider the time dependence of the evolution kernel $\mathcal{E}$. Beginning with Eq. (24), one derives the equation:

$$i\partial_\tau \mathcal{E}(\pi', \pi; \tau) = -\int \mathrm{d}\varphi \, \exp\left(i\pi' \cdot \varphi(\tau) - \pi \cdot \varphi\right) \pi' \cdot (\partial_\tau \varphi(\tau))$$

$$= -\int d\varphi \, \exp\left(i\pi' \cdot \varphi(\tau) - \pi \cdot \varphi\right) \pi' \cdot \omega \cdot \frac{\partial}{\partial \varphi} H(\varphi(\tau))$$

$$= \widehat{\mathcal{H}}(\pi', -i\partial_{\pi'}) \mathcal{E}(\pi', \pi; \tau) , \qquad (25)$$

with the effective *Hamilton operator*:

$$\widehat{\mathcal{H}}(\pi, -i\partial_\pi) \equiv -\pi \cdot \omega \cdot \frac{\partial}{\partial \varphi} H(\varphi)\big|_{\varphi = -i\partial_\pi} . \qquad (26)$$

Here we also used Eq. (22) in the first step, the equations of motion (2) in the second, and suitably pulled the factor following the exponential out of the integral at last. The initial condition,

$$\mathcal{E}(\pi', \pi; 0) = (2\pi)^{2n} \delta^{2n}(\pi' - \pi) , \qquad (27)$$

can be read off from Eq. (23). Integrating Eq. (25), we immediately obtain:

$$\mathcal{E}(\pi', \pi; \tau) = (2\pi)^{2n} \exp[-i\tau \widehat{\mathcal{H}}(\pi', -i\partial_{\pi'})] \delta^{2n}(\pi' - \pi) , \qquad (28)$$

taking the initial condition into account.

This result, which yields the evolution operator $\widehat{U}$ by Eq. (24), finally allows to describe the evolution of a generic time dependent state, $|\Psi(t)\rangle$, which takes place in one discrete physical time step (unit T). Using Eq. (15), we calculate:

$$\langle \tau', \pi' | \Psi(t+T) \rangle = \int d\tau d\pi \, P(\tau) \mathcal{E}(\pi', \pi; \tau' + T - \tau) \langle \tau, \pi | \Psi(t) \rangle$$

$$= \int d\tau \, P(\tau) \exp[-i(\tau' + T - \tau) \widehat{\mathcal{H}}(\pi', -i\partial_{\pi'})] \langle \tau, \pi' | \Psi(t) \rangle \quad (29)$$

where the integration over $d\pi \equiv \prod_a (d\pi_a/2\pi)$ has been carried out with the help of the δ-functions of Eq. (28). This equation appears like a *discrete time Schrödinger equation* and presents our main result of this section. However, some qualifying remarks are due here.

4.1 Discussion

In order to facilitate the investigation of the properties of Eq. (29), let us assume that a generic time dependent state shares the following (proper) time translation property with the state defined in Eq. (11):

$$\langle \tau, \pi | \Psi(t) \rangle = \langle \tau + t, \pi | \Psi(0) \rangle = \langle 0, \pi | \Psi(t + \tau) \rangle . \qquad (30)$$

This is the case, for example, if $|\Psi(t)\rangle$ is defined like $|t\rangle$ there, however, with some particular initial conditions for the paths contributing to the functional integral, or if it is a superposition of such states.

Using this property and shifting and renaming variables, the Eq. (29) can be rewritten:

$$\langle 0, \pi' | \Psi(\tau' + T) \rangle = \int d\tau \, P(\tau - t) \exp[-i(\tau' + T - \tau) \widehat{\mathcal{H}}(\pi', -i\partial_{\pi'})] \langle 0, \pi' | \Psi(\tau) \rangle . \quad (31)$$

Despite that a state needs to exist only on discrete values t of the physical time, the Eqs. (30) or (31) require the corresponding "wave function" in τ, π-representation to be analytically continued to arbitrary real values of the time argument.

Considering the deterministic limiting case, $P(\tau - t) = \delta(\tau - t)$ we obtain directly from Eq. (31):

$$\langle 0, \pi'|\Psi(\tau' + T)\rangle = \exp[-i(\tau' + T - t)\widehat{\mathcal{H}}(\pi', -i\partial_{\pi'})]\langle 0, \pi'|\Psi(t)\rangle \ . \tag{32}$$

Thus, the usual formal solution in terms of an exponentiated Hamilton operator of a standard Schrödinger equation is recovered, i.e., the Eq. (29) here is equivalent to such an equation.

In more general cases, for example, with a Gaussian probability distribution, $P(\tau - t) \propto \exp[-\gamma(\tau - t)^2]$, one might suspect that the evolution operator on the right-hand side of Eq. (29) or (31) is not unitary, even if $\widehat{\mathcal{H}}$ is hermitean. However, it is easily seen that the Ansatz,

$$\langle \tau, \pi|\Psi(t)\rangle \equiv \exp[-i(\tau + t)\widehat{\mathcal{H}}(\pi, -i\partial_\pi)]\langle 0, \pi|\Psi(0)\rangle \ , \tag{33}$$

solves Eq. (29) also for any normalized distribution with $P(\tau; t) = P(\tau - t)$. In this sense, the Eq. (29) is indeed formally equivalent to a Schrödinger equation. We will study this in more detail in Section 6, paying attention to the nonstandard form of the effective Hamiltonian (26) in several examples.

The above Ansatz suggests to introduce *stationary states* defined by the following relations:

$$\langle \tau, \pi|\Psi_n(t)\rangle \equiv \exp(-iE_n t)\langle \tau, \pi|\Psi_n(0)\rangle = \exp(-iE_n(t + \tau))\langle 0, \pi|\Psi_n(0)\rangle \tag{34}$$

$$\equiv \exp(-iE_n(t + \tau))\langle \pi|\Psi_n\rangle \ , \tag{35}$$

in accordance with Eqs. (30), and assuming:

$$\widehat{\mathcal{H}}(\pi, -i\partial_\pi)]\langle 0, \pi|\Psi_n\rangle = E_n\langle 0, \pi|\Psi_n\rangle \ , \tag{36}$$

with a discrete spectrum $\{E_n\}$, in order to be definite.

Finally, we remark that there is no $\hbar$ in our equations. If introduced, it would merely act as a conversion factor of units. On the other hand, there is the intrinsic scale of the clock's unit time interval T. The significance of this can be analyzed in a treatment where clock and system are part of one universe and interact [19].

Before we will illustrate in some examples the type of Hamiltonians and stationary "wave equations" that one obtains, we have to first address the classical observables and their place in the emergent quantum theory. In particular, we need to implement the classical Hamiltonian constraint, which is an essential ingredient related to the gauge symmetry in a time-reparametrization invariant classical theory.

5 Observables

It follows from our introduction of states in Section 3, see particularly Eqs. (8)–(12), how the classical observables of the underlying mechanical system can be calculated. Considering observables which are function(al)s of the phase space variables φ, the definition of their expectation value at physical time t is obvious:

$$\langle O[\varphi]; t\rangle \equiv \int \mathrm{d}\tau \, P(\tau; t) O[-i\frac{\delta}{\delta J(\tau)}] \log Z[J]|_{J=0} \tag{37}$$

$$= Z^{-1} \int d\tau d\pi \, P(\tau - t) \langle 0 | \tau, \pi \rangle O[\varphi(\tau)] \langle \tau, \pi | 0 \rangle \tag{38}$$

$$= Z^{-1} \int d\tau d\pi \, P(\tau) \langle t | \tau, \pi \rangle O[\varphi(\tau + t)] \langle \tau, \pi | t \rangle \tag{39}$$

$$= Z^{-1} \int d\tau d\pi \, P(\tau) \langle t | \tau, \pi \rangle O[-i\partial_\pi] \langle \tau, \pi | t \rangle \tag{40}$$

$$= \langle \Psi(t) | \hat{O}[\varphi] | \Psi(t) \rangle , \tag{41}$$

where $Z \equiv Z[0]$, all states refer to $J = 0$ as well, and where:

$$\hat{O}[\varphi] \equiv O[\hat{\varphi}] , \quad \hat{\varphi} \equiv -i\partial_\pi , \tag{42}$$

in τ, π-representation. In Eqs. (38)–(39) the notation is symbolical, since the observable should be properly included in the functional integral defining the ket state, for example. Furthermore, in the last Eq. (41), the preceding expression is rewritten for the case of a generic state $|\Psi(t)\rangle$ ($J = 0$), with the scalar product to be evaluated as in (40), or as defined in Eq. (14) before.

Thus, a classical observable is represented by the corresponding function(al) of a suitably defined *momentum* operator. Furthermore, its expectation value at physical time t is represented by the effective quantum mechanical expectation value of the corresponding operator with respect to the physical-time dependent state under consideration, which incorporates the weighted average over the proper times τ, according to the distribution P. Not quite surprisingly, the evaluation of expectation values involves an integration over the whole τ-parametrized "history" of the states.

Furthermore, making use of the evolution operator $\hat{U}$ of Section 4, in order to refer observables at different proper times τ_k to a common reference point τ, one can construct *correlation functions* of observables as well, similarly as in Ref. [4], for example.

The most important observable for our present purposes is the classical Hamiltonian, $H(\varphi)$, which enters the Hamiltonian constraint of a classical reparametrization invariant system. It is, by assumption, a constant of the classical motion. However, it is easy to see that also its quantum descendant, $\hat{H}(\varphi) \equiv H(\hat{\varphi})$, is conserved, since it commutes with the effective Hamiltonian of Eq. (26):

$$[\hat{H}, \hat{\mathcal{H}}] = H(-i\partial_\pi) \, \pi \cdot \omega \cdot \frac{\partial}{\partial \varphi} H(\varphi)|_{\varphi = -i\partial_\pi} - \pi \cdot \omega \cdot \frac{\partial}{\partial \varphi} H(\varphi)|_{\varphi = -i\partial_\pi} H(-i\partial_\pi)$$

$$= \frac{\partial}{\partial \varphi} H(\varphi)|_{\varphi = -i\partial_\pi} \cdot \omega \cdot \frac{\partial}{\partial \varphi} H(\varphi)|_{\varphi = -i\partial_\pi} = 0 , \tag{43}$$

due to the antisymmetric character of the symplectic matrix. Therefore, it suffices to implement the Hamiltonian constraint at an arbitrary time.

Then, the constraint of the form $C_H \equiv H(\varphi) - \epsilon \simeq 0$ could be incorporated into the definition of the states in Eq. (11) by including an extra factor $\delta(C_H)$ into the functional integral, and analogously for the adjoint states. Exponentiating the δ-function, we can pull the exponential out of the functional integral, similarly as before. Thus, we find the following operator representing the constraint:

$$\hat{C} \equiv \int d\lambda \, \exp\left(i\lambda(\hat{H}(\varphi) - \epsilon)\right) = \delta(\hat{C}_H) , \tag{44}$$

which acts on states as a projection operator. It admits in the functional integral representing a state only those paths that obey the constraint; in particular, see Eq. (11), it enforces the constraint at the time $\tau + t$. Similarly, a projector can be included into the definition of the generating functional, Eq. (8), in order to represent the Hamiltonian constraint.

Supplementing Eqs. (37)–(41) by the insertion of the Hamiltonian constraint, the properly constrained expection values of observables should be calculated according to:

$$\langle O[\varphi]; t \rangle_H \equiv \langle \Psi(t) | \hat{O}[\varphi] \hat{C} | \Psi(t) \rangle \; , \qquad (45)$$

which, in general, will deviate from the results of the previous definition.

Finally, also the eigenvalue problem of stationary states, see Eqs. (34)–(36), should be studied in the projected subspace:

$$\hat{\mathcal{H}} \hat{C} | \Psi \rangle = E \hat{C} | \Psi \rangle \; , \qquad (46)$$

to which we shall return in the following section.

6 Examples of quantum systems with underlying deterministic dynamics

The purpose of the following examples is to illustrate how the quantum mechanics works in practice which emerges from various deterministic classical systems along the lines presented in Sections 2–5 before. In particular, we solve the stationary Eq. (36) with the effective Hamiltonian $\hat{\mathcal{H}}$ of Eq. (26).

We shall see, however, that acceptable quantum models with a stable groundstate can only be arrived at in this way, if a regularization of the respective Hamiltonian and subsequently a suitable continuum limit are performed. In a sense, the spectrum of the Hamiltonian (26) is too rich, it admits additional unphysical states, which have to be eliminated. While this approach covers a large number of models, the meaning of and possible restrictions on the ad hoc adopted regularization certainly deserve further study.

Our first example, starting with a "timeless" classical harmonic oscillator, mainly serves to demonstrate that the present general formalism allows to recover results of 't Hooft's cellular automaton model [3]. The second model, employing the parameterized classical relativistic particle, is considered by many as a testing ground for any techniques developed to deal with reparametrization invariant theories in general, like general relativity or string theory. It leads to an interacting quantum model, provided we judiciously choose the arbitrary phases introduced by the regularization.

6.1 Quantum system with classical harmonic oscillator beneath

In principle, all *integrable models* can be presented as collections of harmonic oscillators. Therefore, we begin with the harmonic oscillator of unit mass and of frequency Ω. The action is:

$$S \equiv \int dt \left(\frac{1}{2\lambda} (\partial_t q)^2 - \frac{\lambda}{2} (\Omega^2 q^2 - 2\epsilon) \right) \; , \qquad (47)$$

where λ denotes the arbitrary lapse function, i.e. Lagrange multiplier for the Hamiltonian constraint, and $\epsilon > 0$ is the parameter fixing the energy presented by this constraint.

Introducing the proper time, $\tau \equiv \int dt\, \lambda$, the Hamiltonian equations of motion and Hamiltonian constraint for the oscillator are:

$$\partial_\tau q = p\,, \quad \partial_\tau p = -\Omega^2 q\,, \tag{48}$$

$$\frac{1}{2}(p^2 + \Omega^2 q^2) - \epsilon = 0\,, \tag{49}$$

respectively.

Comparing the general structure of the equations of motion, Eq. (2), with the ones obtained here, we identify the effective Hamilton operator, Eq. (26), while the constraint operator follows from Eq. (44):

$$\widehat{\mathcal{H}} = -(\pi_q \widehat{\varphi}_p - \Omega^2 \pi_p \widehat{\varphi}_q) = -\pi_q(-i\partial_{\pi_p}) + \Omega^2 \pi_p(-i\partial_{\pi_q})\,, \tag{50}$$

$$\widehat{C} = \delta(\widehat{\varphi}_p^2 + \Omega^2 \widehat{\varphi}_q^2 - 2\epsilon) = \delta(\partial_{\pi_p}^2 + \Omega^2 \partial_{\pi_q}^2 + 2\epsilon)\,, \tag{51}$$

respectively. Here we employ the convenient notation $\varphi^a \equiv (\varphi_q; \varphi_p)$, and correspondingly $\pi^a \equiv (\pi_q; \pi_p)$, $\partial_\pi^a \equiv (\partial_{\pi_q}; \partial_{\pi_p})$. Further simplifying this with the help of polar coordinates, $\pi_q \equiv -\Omega\rho\cos\phi$ and $\pi_p \equiv \rho\sin\phi$, we obtain:

$$\widehat{\mathcal{H}} = \Omega \widehat{L}_z = -i\Omega \partial_\phi\,, \tag{52}$$

$$\widehat{C} = \delta(\Delta_2 + 2\epsilon) = \delta(\partial_\rho^2 + \rho^{-1}\partial_\rho + \rho^{-2}\partial_\phi^2 + 2\epsilon)\,, \tag{53}$$

where $\widehat{L}_z$ denotes the z-component of the usual angular momentum operator and Δ_2 the Laplacian in two dimensions.

We observe that the eigenfunctions of the eigenvalue problem posed here factorize into a radial and an angular part. The radial eigenfunction, a cylinder function, is important for the calculation of expectation values of certain operators and the overall normalization of the resulting wave functions. However, it does not influence the most interesting spectrum of the Hamiltonian.

In order to proceed, we discretize the angular derivative, Eq. (52). In the absence of the full angular momentum algebra, we would otherwise encounter a discrete yet unbound spectrum, lacking a groundstate.

While we will mostly choose to work with an asymmetric discretization (Case A), we will here also show the symmetric discretization (Case B), in order to appreciate the differences, if any. In any case, the spectrum should and will turn out to be independent of this choice in the continuum limit.

Case A. Here, the energy eigenvalue problem consists in:

$$\widehat{\mathcal{H}}\psi(\phi_n) = -i(\Omega N/2\pi)\big(\psi(\phi_{n+1}) - \psi(\phi_n)\big) = E\psi(\phi_n)\,, \tag{54}$$

with $\phi_n \equiv 2\pi n/N$, $1 \leq n \leq N$, and the continuum limit will be considered momentarily.

The complete orthonormal set of eigenfunctions and the eigenvalues are easily found:

$$\psi_m(\phi_n) = N^{-1/2}\exp[i(m+\delta)\phi_n] \,, \quad 1 \leq m \leq N \,, \tag{55}$$

$$E_m = i(\Omega N/2\pi)\Big(1 - \exp[2\pi i(m+\delta)/N]\Big) \tag{56}$$

$$\stackrel{N\to\infty}{\longrightarrow} \Omega(m+\delta) \,, \quad m \in \mathbf{N} \,, \tag{57}$$

where δ is an arbitrary real constant. Next, we turn to the symmetric discretization.

Case B. Here we have instead:

$$\hat{\mathcal{H}}\psi(\phi_n) = -i(\Omega N/4\pi)\Big(\psi(\phi_{n+1}) - \psi(\phi_{n-1})\Big) = E\psi(\phi_n) \,, \tag{58}$$

with ϕ_n as before. This is solved by the same eigenfunctions as before, Eq. (55). However, in this case the eigenvalues are real even before taking the continuum limit:

$$E_m = (\Omega N/2\pi)\sin[2\pi(m+\delta)/N] \stackrel{N\to\infty}{\longrightarrow} \Omega(m+\delta) \,, \quad m \in \mathbf{N} \,. \tag{59}$$

Thus, we learn that the spectrum in the continuum limit is indeed real and independent of the regularizations employed to suitably define the Hamiltonian, as it should be.

The freedom to choose the constant δ, which arises from the regularization of the Hamilton operator, is most wellcome. Choosing $\delta \equiv -1/2$, we arrive at the *quantum harmonic oscillator*, having started from a corresponding classical system here. Thus, we recover in a straightforward way 't Hooft's result, derived from a cellular automaton [3]. See also Ref. [9] for the completion of a similar quantum model. In the following example we will encounter one more model of this kind and demonstrate its solution in detail.

Also in the following example the regularized eigenvalues are *complex* and the real spectrum only emerges in the continuum limit. Again, this is due to the fact that we choose to discretize first-order derivatives asymmetrically, and could be avoided as shown.

6.2 Quantum system with classical relativistic particle beneath

Introducing proper time similarly as in Ref. [9], the equations of motion and the Hamiltonian constraint of the reparametrization invariant kinematics of a classical relativistic particle of mass m are given by:

$$\partial_\tau q^\mu = m^{-1}p^\mu \,, \quad \partial_\tau p^\mu = 0 \,, \tag{60}$$

$$p \cdot p - m^2 = 0 \,, \tag{61}$$

respectively. Here we have $\varphi^a \equiv (q^0, \ldots, q^3; p^0, \ldots, p^3)$, $a = 1, \ldots, 8$; four-vector products involve the Minkowski metric, $g^{\mu\nu} \equiv \text{diag}(1, -1, -1, -1)$.

Proceeding as before, we identify the effective Hamilton operator:

$$\hat{\mathcal{H}} = -m^{-1}\pi_q \cdot \hat{\varphi}_p = -m^{-1}\pi_q \cdot (-i\partial_{\pi_p}) \,, \tag{62}$$

corresponding to Eq. (26); the notation is as introduced after Eq. (51), however, involving four-vectors. Furthermore, the Hamiltonian constraint is represented by the operator:

$$\hat{C} = \delta(\hat{\varphi}_p^2 - m^2) = \delta(\partial_{\pi_p}^2 + m^2) \,, \tag{63}$$

according to Eqs. (44) and (61). After a Fourier transformation, which replaces the variable π_q by a derivative (four-vector) $+i\partial_x$, and renaming $\pi_p \equiv \bar{x}$, we obtain:

$$\widehat{\mathcal{H}} = -m^{-1}\partial_x \cdot \partial_{\bar{x}} \; , \quad \widehat{C} = \delta(\partial_{\bar{x}}^2 + m^2) \; , \tag{64}$$

i.e., a more transparent form of the Hamilton and constraint operators, respectively.

Before embarking to further analyze this model, some general remarks seem in order here. It is well known from ordinary quantum (field) theory that the harmonic oscillator is peculiar in many respects. Therefore, the reader should not be misled by the results of Section 6.1, where we obtain a quantum harmonic oscillator spectrum from an underlying classical harmonic oscillator model. In particular, it may appear as if we have invented just one more quantization method, in line with quantization via canonical commutators, stochastic quantization, etc. However, we stress that this is *not* the case.

It seems an accident of the harmonic system that the usual quantized energy spectrum results here. This is revealed by the fact, already demonstrated in Refs. [3, 8, 9], that localization with respect to the coordinate q of the underlying classical model has nothing to do with localization with respect to the operator $\hat{q}$, which is introduced a posteriori when interpreting the emergent quantum Hamiltonian corresponding to said spectrum. Rather, such localized quantum (oscillator) states are widely spread over the q-space of the underlying classical model. We will encounter such operators in the following example again. Furthermore, the usual $\hat{p}, \hat{q}$-commutator algebra obtains corrections here, as long as the regularization is not removed.

Therefore, generally, we do not expect to find the usual quantized counterpart of a classical reparametrization invariant model in the present approach based on discrete physical time. There will not be the usual one-to-one correspondence. To put it differently, the *classical limit* of emergent quantum theories is not at all expected to give back the *underlying* classical model. Further general remarks in this vein may be found in Ref. [3].

The following discussion of the free relativistic particle should be seen in this light. While any standard quantization method produces an unbound spectrum with notorious negative energy states, we show in detail here that careful application of the freedom introduced by the regularization does indeed produce an acceptable quantum model in the continuum limit, which is very different from the underlying classical model.

The eigenvalue problem is solved after discretizing the system with a hypercubic (phase space) lattice of volume L^8 (lattice spacing $l \equiv L/N$) and periodic boundary conditions, for example. Similarly as in the oscillator case, we here obtain the eigenfunctions:

$$\psi_{k_x,k_{\bar{x}}}(x_n, \bar{x}_n) = N^{-1} \exp[i(k_x + \delta_x) \cdot x_n + i(k_{\bar{x}} + \delta_{\bar{x}}) \cdot \bar{x}_n] \; , \tag{65}$$

with coordinates $x_n^\mu \equiv l n^\mu$ and momenta $k_x^\mu \equiv 2\pi k^\mu/L$, with $1 \leq n^\mu, k^\mu \leq N$, and where δ_x^μ are arbitrary real constants, for all $\mu = 0, \ldots, 3$ (analogously $\bar{x}_n^\mu, k_{\bar{x}}^\mu, \delta_{\bar{x}}^\mu$).

These are the eigenfunctions, which are of plane-wave type, of the stationary Schrödinger equation, $\widehat{\mathcal{H}}\widehat{C}\psi = E\widehat{C}\hat{\psi}$, with $\widehat{\mathcal{H}}$ and $\widehat{C}$ from Eqs. (64) discretized analogously to Case A of Section 6.1. The four-vector momenta $k_x^\mu, k_{\bar{x}}^\mu$ label different eigenfunctions, a familiar feature in quantum mechanics, and the constraint will be implemented shortly. Note that the phases $\delta_x^\mu, \delta_{\bar{x}}^\mu$ can still be chosen arbitrarily; however, since they must be fixed once for all, they cannot possibly absorb the variable four-momenta. Furthermore, we emphasize that the phase space lattice has been introduced for the only purpose of regularizing the Hamiltonian, to be followed by a suitable continuum limit. At present, however, it is unrelated to the underlying discreteness of physical time.

The corresponding energy eigenvalues are:

$$E_{k_x,k_{\bar{x}}} = -m^{-1}l^{-2}\Big((\exp[il(k_x+\delta_x)^0]-1)(\exp[il(k_{\bar{x}}+\delta_{\bar{x}})^0]-1) \quad (66)$$

$$-\sum_{j=1}^{3}(\exp[il(k_x+\delta_x)^j]-1)(\exp[il(k_{\bar{x}}+\delta_{\bar{x}})^j]-1)\Big)$$

$$= m^{-1}(k_x+\delta_x)\cdot(k_{\bar{x}}+\delta_{\bar{x}}) + O(l) , \quad (67)$$

where is L is kept constant in the continuum limit, $l \to 0$; again, the four-momenta $k_x^\mu, k_{\bar{x}}^\mu$ label and determine the energies. Furthermore, in this limit, one finds that the Hamiltonian constraint requires timelike "on-shell" vectors $k_{\bar{x}}$, obeying $(k_{\bar{x}}+\delta_{\bar{x}})^2 = m^2$, while leaving k_x unconstrained.

Continuing, we perform also the infinite volume limit, $L \to \infty$, which results in a continuous energy spectrum in Eq. (67). We observe that no matter how we choose the constants $\delta_x, \delta_{\bar{x}}$, the spectrum will not be positive definite. Thus, the emergent model appears not to be acceptable, since it does not lead to a stable groundstate.

However, let us proceed more carefully with the various limits involved and show that indeed a well-defined quantum model is obtained. For simplicity, considering (1+1)-dimensional Minkowski space and anticipating the massless limit, we rewrite Eq. (67) explicitly:

$$E_{k,\bar{k}} = -(\frac{2\pi}{\sqrt{m}L})^2(\bar{k}^1+\bar{\delta}^1)\Big((k^0+\delta^0)+(k^1+\delta^1)\Big) + O(m) , \quad (68)$$

where we suitably rescaled and renamed the constants and the momenta, which run in the range $1 \leq \bar{k}^1, k^{0,1} \leq N \equiv 2s+1$. Furthermore, we incorporated the Hamiltonian (on-shell) constraint, such that only the positive root contributes: $\bar{k}^0 + \bar{\delta}^0 = |\bar{k}^1 + \bar{\delta}^1| + O(m^2) = -(\bar{k}^1 + \bar{\delta}^1) + O(m^2)$. This can be achieved by suitably choosing $\bar{\delta}^{0,1}$.

In fact, just as in the previous harmonic oscillator case, the choice of the phase constants is crucial in determining the quantum model. Here we set:

$$\bar{\delta}^0 \equiv \frac{1}{2}, \quad \bar{\delta}^1 \equiv \frac{1}{2} - 2s - 3 , \quad \delta^{0,1} \equiv 0 . \quad (69)$$

We observe that this choice implies that in the continuum limit, with $l \to 0$, since $N \to \infty$, we have $s \to \infty$ and thus $\bar{\delta}^1 \to -\infty$. Incorporating these phases, we obtain the manifestly positive definite spectrum:

$$E(\bar{s}_z, s_z^{0,1}) = (\frac{2\pi}{\sqrt{m}L})^2\Big((\bar{s}_z+s+\frac{1}{2})+1\Big)\Big((s_z^0+s+\frac{1}{2})+(s_z^1+s+\frac{1}{2})+1\Big)+O(m) , \quad (70)$$

where also the (half)integer quantum numbers $\bar{s}_z, s_z^{0,1}$ are introduced, all in the range $-s \leq s_z \leq s$, which correctly replace $\bar{k}^1, k^{0,1}$.

Recalling the algebra of the $SU(2)$ generators, with $S_z|s_z\rangle = s_z|s_z\rangle$ in particular, we are led to consider the generic operator:

$$h \equiv S_z + s + \frac{1}{2} \ , \tag{71}$$

i.e., diagonal with respect to $|s_z\rangle$-states of the (half)integer representations determined by s. In terms of such operators, we obtain the regularized Hamiltonian corresponding to Eq. (70):

$$\widehat{\mathcal{H}} = (\frac{2\pi}{\sqrt{m}L})^2 \Big(1 + \bar{h} + h_0 + h_1 + \bar{h}(h_0 + h_1)\Big) + O(m) \ , \tag{72}$$

which will turn out to represent three coupled harmonic oscillators, including an additional contribution to the vacuum energy.

A Hamiltonian of the type of h has been the starting point of 't Hooft's analysis [3], which we adapt for our purposes in the following.

Continuing with standard notation, we have $S^2 \equiv S_x^2 + S_y^2 + S_z^2 = s(s+1)$, which suffices to obtain the following identity:

$$h = \frac{1}{2s+1}\Big(S_x^2 + S_y^2 + \frac{1}{4} + h^2\Big) \ . \tag{73}$$

Furthermore, using $S_\pm \equiv S_x \pm iS_y$, we define coordinate and conjugate momentum operators:

$$\hat{q} \equiv \frac{1}{2}(aS_- + a^*S_+) \ , \quad \hat{p} \equiv \frac{1}{2}(bS_- + b^*S_+) \ , \tag{74}$$

where a and b are complex coefficients. Calculating the basic commutator with the help of $[S_+, S_-] = 2S_z$ and using Eq. (71), we obtain:

$$[\hat{q}, \hat{p}] = i(1 - \frac{2}{2s+1}h) \ , \tag{75}$$

provided we set $\Im(a^*b) \equiv -2/(2s+1)$. Incorporating this, we calculate:

$$S_x^2 + S_y^2 = \frac{(2s+1)^2}{4}\Big(|a|^2\hat{p}^2 + |b|^2\hat{q}^2 - (\Im a \cdot \Im b + \Re a \cdot \Re b)\{\hat{q}, \hat{p}\}\Big) \ . \tag{76}$$

In order to obtain a reasonable Hamiltonian in the continuum limit, we set:

$$a \equiv i\frac{\Omega^{-1/2}}{\sqrt{s+1/2}} \ , \quad b \equiv \frac{\Omega^{1/2}}{\sqrt{s+1/2}} \ , \quad \Omega \equiv (\frac{2\pi}{\sqrt{m}L})^2 \ . \tag{77}$$

Then, the previous Eq. (73) becomes:

$$\Omega h = \frac{1}{2}\hat{p}^2 + \frac{1}{2}\Omega^2\hat{q}^2 + \frac{1}{(2s+1)\Omega}\Big(\frac{1}{4}\Omega^2 + (\Omega h)^2\Big) \ , \tag{78}$$

reveiling a nonlinearly modified harmonic oscillator Hamiltonian, similarly as in Ref. [9].

Now it is safe to consider the continuum limit, $2s + 1 = N \to \infty$, keeping $\sqrt{m}L$ and Ω finite. This produces the usual $\hat{q}, \hat{p}$-commutator in Eq. (75) for states with limited energy and the standard harmonic oscillator Hamiltonian in Eq. (78).

Using these results in Eq. (72), the Hamilton operator of the emergent quantum model is obtained:

$$\hat{\mathcal{H}} = \Omega + \frac{1}{2} \sum_{j=\bar{1},0,1} \left(\hat{p}_j^2 + \Omega^2 \hat{q}_j^2 \right) + \frac{1}{4\Omega}(\hat{p}_{\bar{1}}^2 + \Omega^2 \hat{q}_{\bar{1}}^2) \sum_{j=0,1} \left(\hat{p}_j^2 + \Omega^2 \hat{q}_j^2 \right) , \qquad (79)$$

where the massless limit together with the infinite volume limit is carried out, $m \to 0$, $L \to \infty$, in such a way that Ω remains finite.

The resulting Hamiltonian here is well defined in terms of continuous operators $\hat{q}$ and $\hat{p}$, as usual, and has a positive definite spectrum. The coupling term might appear somewhat more familiar, if the oscillator algebra is realized in terms of bosonic creation and annihilation operators.

In Refs. [8, 9] we calculated the matrix elements of operators $\hat{q}, \hat{p}$ with respect to the SU(2) basis of primordial states in an analogous case, showing that *localization of the quantum oscillator has little to do with localization in the classical model beneath*, as we mentioned before.

Finally, we remark that had we chosen $\bar{\delta}^{0,1} = \delta^{0,1} \equiv 1/2$, instead of Eqs. (69), then a relative sign between terms would remain, originating from the Minkowski metric, and this would yield the Hamiltonian $\hat{\mathcal{H}} \propto (1 + \bar{h})(h_0 - h_1)$, which is not positive definite. Similarly, any symmetric choice, $\bar{\delta}^{0,1} = \delta^{0,1} \equiv \delta$ would suffer from this problem.

This raises the important issue of the role of canonical transformations, and of symmetries in particular. It is conceivable that symmetries will play a role in restricting the apparent arbitrariness of the regularization. Independence from the choice of phase space coordinates employed in actual calculations might be a desirable feature. A preliminary study of an anharmonic oscillator indicates that the simple discretization used here possibly needs to be improved, in order to fullfill this.

6.3 Remarks

Concluding this section, we may state that the features which we illustrated in the previous examples, especially the necessity of regularization (discretization), promise to make genuinely interacting models quite difficult to analyze. Interesting results may perhaps be found with the help of spectrum generating algebras and/or some to-be-developed perturbative methods.

We find it interesting that our general Hamilton operator, Eq. (26), does not allow for the direct addition of a constant energy term, while in its regularized form this is possible, due to the appearance of an arbitrary phase. Choosing the latter determines the groundstate energy, which survives the continuum limit.

Leaving aside the Hamiltonian constraint momentarily, we observe that the Hamilton operator equation, also in cases with large numbers n of coupled degrees of freedom, amounts to systems of first order quasi-linear partial differential equations. They can be studied by the method of characteristics [23]. Thus, one finds one inhomogeneous equation, which can be trivially integrated. Furthermore, the remaining $2n$ equations for the characteristics present nothing but the classical Hamiltonian equations of motion.

Therefore, integrable classical models can be decoupled at the level of the characteristic equations by canonical transformations, if they can be applied freely at the pre-quantum level. Classical crystal-like models with harmonic forces, or free field theories, respectively, will thus give rise to corresponding free quantum mechanical systems. These are constructed in a different way in Refs. [3]. Presumably, the (fixing of a large class of) gauge transformations invoked there can be related to the existence of integrals of motion implied by integrability here.

Finally, we point out that the Hamiltonian equations of motion preclude motion into classically forbidden regions of the underlying system. Nevertheless, *quantum mechanical tunneling* is an intrinsic property of the quantum oscillator models that we obtained.

7 Conclusions

In the present chapter, we pursue the view that quantum mechanics is an emergent description of nature, which possibly can be based on classical, pre-quantum concepts.

Our approach is motivated by the ongoing construction of a reparametrization-invariant time. In turn, this is based on the observation that "time passes" when there is an observable change, which is localized with the observer. More precisely, necessary are incidents, i.e. observable unit changes, which are recorded, and from which invariant quantities characterizing the change of the evolving system can be derived [8, 9].

Presently, this has led us to assume the relation between the constructed physical time t and standard proper time τ of the evolving system in the form of a statistical distribution, $P(\tau; t) = P(\tau - t)$, cf. Eq. (7). We assume that the distribution is not explicitly time-dependent, which means, the physical clock is practically decoupled from the system under study. We explore the consequences of this situation for the description of the system.

We have shown how to introduce "states", eventually building up a Hilbert space, in terms of certain functional integrals, Eqs. (11)–(12), which arise from the study of a classical generating functional. The latter was introduced earlier in a different context, studying classical mechanics in functional form [21, 22]. We employ this as a convenient tool, and modify it, in order to describe the observables of reparametrization-invariant systems with discrete time (Section 5). Studying the evolution of the states in general (Section 4), we are led to the Schrödinger equation, Eq. (29). However, the Hamilton operator, Eq. (26), has a non-standard form.

We demonstrate that proper regularization of the continuum Hamilton operator is indispensable, in order to find a stable groundstate. Limitations imposed by symmetries or further constraints and consistency of the procedure need to be clarified. Other possible regularization schemes need to be explored.

Coming back to the probabilistic relation between physical time and the evolution parameter figuring in the parameterized classical equations of motion: One would like to include the clock degrees of freedom consistently into the dynamics, in order to address the closed universe. This can be achieved by introducing suitable projectors into the generating functional [13, 19]. Their task is to represent quasi-local detectors which respond to a particle trajectory passing through in Yes/No fashion. In a more general setting, such a detector/projector has to be defined in terms of observables of the closed system. In this way, typical conditional probabilities can be handled, such as describing "What is the probability of observable X having a value in a range x to $x + \delta x$, *when* observable Y has value y?". Criteria for selecting the to-be-clock degrees of freedom are still unknown, other than simplicity. Most likely the resulting description of evolution and implicit notion of physical time will correspond to our distribution $P(\tau;t)$ of Eq. (7), however, evolving with the system. Dissipative and memory effects, which will arise after integration over clock degrees of freedom, might play a crucial role in deriving a unique large-scale quantum model.

This stroboscopic quantization emerging from underlying classical dynamics certainly may be questioned in many respects. It might violate one or the other assumption of existing no-go theorems relating to hidden variables theories. However, we believe it is interesting to get closer to a working example, before discussing this. Unitary evolution and tunneling effects are recovered in this framework, under the proviso of regularization of the continuum formalism.

Acknowledgements

It is a pleasure to thank Christof Wetterich for discussion and the members of the Institut für Theoretische Physik (Heidelberg) for their kind hospitality, while this work was begun. Correspondence from Ennio Gozzi is thankfully acknowledged. This work has been supported by CNPq (Brasilia) 690138/02-4 and DAAD (Bonn) A/03/17806.

Note added in proof

Since the time of submission of this article, I have pursued the question of how to construct a stable groundstate for emergent quantum models (or handle the possible instability) in several further examples [24]. In particular, various field theories have been considered; a special class are supersymmetric models, were the additional symmetry is employed to resolve these issues.

References

[1] A. Einstein, B. Podolsky and N. Rosen, *Phys. Rev.* **47**, 777 (1935).

[2] J.S. Bell, ¿*Speakable and Unspeakable in Quantum Mechanics*î (Cambridge University Press, Cambridge, 1987).

[3] G. 't Hooft, *Quantum Mechanics and Determinism*, in: Proceedings of the Eigth Int. Conf. on "Particles, Strings and Cosmology", P. Frampton and J. Ng, Eds. (Rinton Press, Princeton, 2001), p.275; [hep-th/0105105]; see also: *Determinism Beneath Quantum Mechanics*, [quant-ph/0212095].

[4] C. Wetterich, *Quantum correlations in classical statistics*, [quant-ph/0212031], in: ¿*Decoherence and Entropy in Complex Systems*î, H.-T. Elze (ed.), Lecture Notes in Physics, Vol. 633 (Springer-Verlag, Berlin Heidelberg New York, 2003).

[5] M. Blasone, P. Jizba and G. Vitiello, *Phys. Lett.* **A287**, 205 (2001); *Dissipation, Emergent Quantization and Quantum Fluctuations*, [quant-ph/0301031], in: see [4].

[6] F. Markopoulou and L. Smolin, *Phys. Rev. D* **70**, 124029 (2004).

[7] S.L. Adler, *Quantum Mechanics as an Emergent Phenomenon* (Cambridge U. Press, Cambridge, 2005).

[8] H.-T. Elze and O. Schipper, *Phys. Rev. D* **66**, 044020 (2002).

[9] H.-T. Elze, *Phys. Lett. A* **310**, 110 (2003).

[10] C. Rovelli, *Phys. Rev. D* **42**, 2638 (1990).

[11] M. Montesinos, C. Rovelli and Th. Thiemann, *Phys. Rev. D* **60**, 044009 (1999).

[12] M. Montesinos, *Gen. Rel. Grav.* **33**, 1 (2001).

[13] J.J. Halliwell and J. Thorwart, *Phys. Rev. D* **65**, 104009 (2002).

[14] H.S. Snyder, *Phys. Rev.* **71**, 38 (1947).

[15] T.D. Lee, *Phys. Lett.* **122B**, 217 (1983).

[16] G. Jaroszkiewicz and K. Norton, *J. Phys. A* **30**, 3115 (1997); *A* **30**, 3145 (1997); *A* **31**, 977 (1998).

[17] R. Gambini and J. Pullin: *Phys. Rev. Lett.* **90**, 021301 (2003); *'Discrete quantum gravity: applications to cosmology'*, gr-qc/0212033; C. Di Bartolo, R. Gambini and J. Pullin: *Class. Quant. Grav.* **19**, 5275 (2002).

[18] G. Amelino-Camelia, *The three perspectives on the quantum-gravity problem and their implications for the fate of Lorentz symmetry*, based on invited seminars at "Perspectives on Quantum Gravity: a tribute to John Stachel" (Boston, March 6-7, 2003) and "Tenth Marcel Grossmann Meeting on General Relativity" (Rio de Janeiro, July 20-26, 2003), [gr-qc/0309054].

[19] H.-T. Elze, in preparation.

[20] B.O. Koopman, *Proc. Nat. Acad. Sci.* (USA) **17**, 315 (1931); J. von Neumann, *Ann. Math.* **33**, 587 (1932).

[21] E. Gozzi, M. Reuter and W.D. Thacker, *Phys. Rev. D* **40**, 3363 (1989); E. Gozzi, M. Reuter and W.D. Thacker, *Phys. Rev. D* **46**, 757 (1992).

[22] E. Gozzi and M. Regini, *Phys. Rev. D* **62**, 067702 (2000); A.A. Abrikosov (jr.) and E. Gozzi, *Nucl. Phys. Proc. Suppl.* **66**, 369 (2000).

[23] R. Courant and D Hilbert, *Methods Of Mathematical Physics*, Vol. II (Interscience Publ., New York, 1962).

[24] H.-T. Elze, *Quantum fields, cosmological constant and symmetry doubling*, arXiv: hep-th/0510267; *Braz. J. Phys.* **35**, 343 (2005); *A quantum field theory as emergent description of constrained supersymmetric classical dynamics*, Proceedings 8th Int. Conf. *Path Integrals. From Quantum Information to Cosmology*, Prague, June 6-10, 2005, to appear, arXiv: hep-th/0508095; *Phys. Lett. A* **335**, 258 (2005); *Physica A* **344**, 478 (2004).

Chapter 6

CANONICAL STRUCTURE OF 3D GRAVITY WITH TORSION

M. Blagojević and B. Cvetković
Institute of Physics,
P. O. Box 57, 11001 Belgrade, Serbia
Email addresses: mb@phy.bg.ac.yu, cbranislav@phy.bg.ac.yu

Abstract

We study the canonical structure of the topological 3D gravity with torsion, assuming the anti-de Sitter asymptotic conditions. It is shown that the Poisson bracket algebra of the canonical generators has the form of two independent Virasoro algebras with classical central charges. In contrast to the case of general relativity with a cosmological constant, the values of the central charges are different from each other.

1 Introduction

Faced with enormous difficulties to properly understand fundamental dynamical properties of gravity, such as the nature of classical singularities and the problem of quantization, one is naturally led to consider technically simplified models with the same conceptual features. An important and useful model of this type is 3D gravity [1, 2]. In the last twenty years, 3D gravity has become an active research area, with a number of outstanding results. Here, we focus our attention on a particular line of development, characterized by the following achievements. In 1986, Brown and Henneaux introduced the so-called anti-de Sitter (AdS) asymptotic conditions in their study of 3D general relativity with a cosmological constant (GR_Λ) [3]. They showed that the related behavior of the gravitational field allows for an extremely rich asymptotic structure—the conformal symmetry described by two independent canonical Virasoro algebras with classical central charges. Soon after that, Witten rediscovered and further explored the fact that GR_Λ in 3D can be formulated as a Chern-Simons gauge theory [4]. The equivalence between gravity and an ordinary gauge theory was shown to be crucial for our understanding of quantum gravity. Then, in 1993, we had the discovery of the BTZ black hole [5], with a far-reaching impact on the development of 3D gravity. All these ideas have had a significant influence on our understanding of the quantum nature of 3D black holes [2, 6–13].

Following a widely spread belief that general relativity is the most reliable approach for studying the gravitational phenomena, the analysis of these issues has been carried out mostly in the realm of *Riemannian* geometry. However, there is a more general conception of gravity based on *Riemann-Cartan* geometry, in which both the curvature and the torsion characterize the structure of gravity (see, for instance, Refs. [14, 15]). Riemann-Cartan geometry has been used in the context of 3D gravity since the early 1990s [16–18], with an idea to explore the influence of geometry on the dynamics of gravity. Recently, new advances in this direction have been achieved [19–24].

Asymptotic conditions are an intrinsic part of the canonical formalism, as they define the phase space in which the canonical dynamics takes place. Their influence on the dynamics is particularly clear in topological theories, where the propagating degrees of freedom are absent, and the only non-trivial dynamics is bound to exist at the asymptotic boundary. General action for topological 3D gravity with torsion, based on Riemann-Cartan geometry of spacetime, has been proposed by Mielke and Baekler [16, 17]. The objective of the present paper is to investigate the canonical structure of the general topological 3D gravity with torsion, including its asymptotic behavior, in the AdS sector of the theory. This will generalize the results obtained in Refs. [3, 4] and [20], where the specific choice of parameters corresponds to Riemannian and teleparallel vacuum geometry, respectively. Combining this approach with another interesting result, the existence of the Riemann-Cartan black hole [19, 22], we shall be able to explore the full influence of torsion on the canonical and asymptotic structure of 3D gravity.

The paper is organized as follows. In Sect. 2 we review some basic features of Riemann–Cartan spacetime as the proper geometric arena for 3D gravity with torsion, and discuss the field equations derived from the Mielke-Baekler action. In Sect. 3 we describe the Riemann-Cartan black hole solution, a generalization of the BTZ black hole. Then, in Sect. 4, we introduce the concept of asymptotically AdS configuration, and derive the related asymptotic symmetry, which turns out to be the same as in general relativity—the conformal symmetry. In the next section, the asymptotic structure of the theory is incorporated into the Hamiltonian formalism by calculating the Poisson bracket (PB) algebra of the canonical generators. It has the form of two independent Virasoro algebras with classical central charges, the values of which differ from each other, in contrast to what we have in Riemannian GR_Λ and the teleparallel theory [3, 20]. Finally, Sect. 7 is devoted to concluding remarks, while Appendices contain some technical details.

Our conventions are given by the following rules: the Latin indices refer to the local Lorentz frame, the Greek indices refer to the coordinate frame; the first letters of both alphabets $(a, b, c, ...; \alpha, \beta, \gamma, ...)$ run over 1,2, the middle alphabet letters $(i, j, k, ...; \mu, \nu, \lambda, ...)$ run over 0,1,2; the triad field $b^i{}_\mu$ and its dual $h_i{}^\mu$ are used to convert Greek and Latin indices into each other; $\eta_{ij} = (+, -, -)$ and $g_{\mu\nu} = \eta_{ij} b^i{}_\mu b^j{}_\nu$ are the metric components in the local Lorentz and coordinate frame; totally antisymmetric tensor ε^{ijk} and the related tensor density $\varepsilon^{\mu\nu\rho}$ are both normalized so that $\varepsilon^{012} = 1$.

2 Topological 3D gravity with torsion

Theory of gravity with torsion can be formulated as Poincaré gauge theory (PGT), with an underlying spacetime structure described by Riemann-Cartan geometry [14, 15].

PGT in brief. The basic gravitational variables in PGT are the triad field b^i and the Lorentz connection $A^{ij} = -A^{ji}$ (1-forms). The field strengths corresponding to the gauge potentials b^i and A^{ij} are the torsion T^i and the curvature R^{ij} (2-forms): $T^i = db^i + A^i{}_m \wedge b^m$, $R^{ij} = dA^{ij} + A^i{}_m \wedge A^{mj}$. Gauge symmetries of the theory are local translations and local Lorentz rotations, parametrized by ξ^μ and ε^{ij}.

In 3D, we can simplify the notation by introducing the duals of A^{ij}, R^{ij} and ε^{ij}:

$$\omega_i = -\frac{1}{2}\varepsilon_{ijk}A^{jk}, \qquad R_i = -\frac{1}{2}\varepsilon_{ijk}R^{jk}, \qquad \theta_i = -\frac{1}{2}\varepsilon_{ijk}\varepsilon^{jk}.$$

In local coordinates x^μ, we can expand the triad and the connection 1-forms as $b^i = b^i{}_\mu dx^\mu$, $\omega^i = \omega^i{}_\mu dx^\mu$. Gauge transformation laws have the form

$$\delta_0 b^i{}_\mu = -\varepsilon^i{}_{jk}b^j{}_\mu \theta^k - (\partial_\mu \xi^\rho)b^i{}_\rho - \xi^\rho \partial_\rho b^i{}_\mu \equiv \delta_{\mathrm{PGT}} b^i{}_\mu,$$
$$\delta_0 \omega^i{}_\mu = -(\partial_\mu \theta^i + \varepsilon^i{}_{jk}\omega^j{}_\mu \theta^k) - (\partial_\mu \xi^\rho)\omega^i{}_\rho - \xi^\rho \partial_\rho \omega^i{}_\mu \equiv \delta_{\mathrm{PGT}} \omega^i{}_\mu, \qquad (2.1)$$

and the field strengths are given as

$$T^i = \nabla b^i \equiv db^i + \varepsilon^i{}_{jk}\omega^j \wedge b^k = \frac{1}{2}T^i{}_{\mu\nu}dx^\mu \wedge dx^\nu,$$
$$R^i = d\omega^i + \frac{1}{2}\varepsilon^i{}_{jk}\omega^j \wedge \omega^k = \frac{1}{2}R^i{}_{\mu\nu}dx^\mu \wedge dx^\nu, \qquad (2.2)$$

where $\nabla = dx^\mu \nabla_\mu$ is the covariant derivative.

To clarify the geometric meaning of the above structure, we introduce the metric tensor as a specific, bilinear combination of the triad fields:

$$g = \eta_{ij}b^i \otimes b^j = g_{\mu\nu}dx^\mu \otimes dx^\nu,$$
$$g_{\mu\nu} = \eta_{ij}b^i{}_\mu b^j{}_\nu, \qquad \eta_{ij} = (+,-,-).$$

Although metric and connection are in general independent dynamical/geometric variables, the antisymmetry of A^{ij} in PGT is equivalent to the so-called *metricity condition*, $\nabla g = 0$. The geometry whose connection is restricted by the metricity condition (metric-compatible connection) is called *Riemann-Cartan geometry*. Thus, PGT has the geometric structure of Riemann-Cartan space.

The connection A^{ij} determines parallel transport in the local Lorentz basis. Being a true geometric operation, parallel transport is independent of the basis. This property is incorporated into PGT via the so-called *vielbein postulate*, which implies the identity

$$A_{ijk} = \Delta_{ijk} + K_{ijk}, \qquad (2.3)$$

where Δ is Riemannian (Levi-Civita) connection, and $K_{ijk} = -\frac{1}{2}(T_{ijk} - T_{kij} + T_{jki})$ is the contortion.

Topological action. In general, gravitational dynamics is defined by Lagrangians which are at most quadratic in field strengths. Omitting the quadratic terms, Mielke and Baekler proposed a *topological* model for 3D gravity [16, 17], with an action of the form

$$I = aI_1 + \Lambda I_2 + \alpha_3 I_3 + \alpha_4 I_4 + I_M, \tag{2.4a}$$

where I_M is a matter contribution, and

$$I_1 = 2\int b^i \wedge R_i,$$
$$I_2 = -\frac{1}{3}\int \varepsilon_{ijk} b^i \wedge b^j \wedge b^k,$$
$$I_3 = \int \left(\omega^i \wedge d\omega_i + \frac{1}{3}\varepsilon_{ijk}\omega^i \wedge \omega^j \wedge \omega^k\right),$$
$$I_4 = \int b^i \wedge T_i. \tag{2.4b}$$

The first term, with $a = 1/16\pi G$, is the usual Einstein-Cartan action, the second term is a cosmological term, I_3 is the Chern-Simons action for the Lorentz connection, and I_4 is an action of the translational Chern-Simons type. The Mielke-Baekler model can be thought of as a natural generalization of Riemannian GR_Λ (with $\alpha_3 = \alpha_4 = 0$) to a topological gravity theory in Riemann-Cartan spacetime.

Field equations. Variation of the action with respect to triad and connection yields the gravitational field equations:

$$\varepsilon^{\mu\nu\rho}\left[aR_{i\nu\rho} + \alpha_4 T_{i\nu\rho} - \Lambda\varepsilon_{ijk}b^j{}_\nu b^k{}_\rho\right] = \tau^\mu{}_i,$$
$$\varepsilon^{\mu\nu\rho}\left[\alpha_3 R_{i\nu\rho} + aT_{i\nu\rho} + \alpha_4\varepsilon_{ijk}b^j{}_\nu b^k{}_\rho\right] = \sigma^\mu{}_i,$$

where $\tau^\mu{}_i = -\delta I_M/\delta b^i{}_\mu$ and $\sigma^\mu{}_i = -\delta I_M/\delta \omega^i{}_\mu$ are the matter energy-momentum and spin currents, respectively. For our purposes—to study the canonical structure of the theory in the asymptotic region—it is sufficient to consider the field equations in vacuum, where $\tau = \sigma = 0$. For $\alpha_3\alpha_4 - a^2 \neq 0$, these equations take the form

$$T_{ijk} = p\varepsilon_{ijk}, \tag{2.5a}$$
$$R_{ijk} = q\varepsilon_{ijk}, \tag{2.5b}$$

where

$$p = \frac{\alpha_3\Lambda + \alpha_4 a}{\alpha_3\alpha_4 - a^2}, \qquad q = -\frac{(\alpha_4)^2 + a\Lambda}{\alpha_3\alpha_4 - a^2}.$$

Thus, the vacuum configuration is characterized by constant torsion and constant curvature.

In Riemann-Cartan spacetime, one can use the identity (2.3) to express the curvature $R^{ij}{}_{\mu\nu}(A)$ in terms of its Riemannian piece $\tilde{R}^{ij}{}_{\mu\nu} \equiv R^{ij}{}_{\mu\nu}(\Delta)$ and the contortion:

$$R^{ij}{}_{\mu\nu}(A) = \tilde{R}^{ij}{}_{\mu\nu} + \left[\nabla_\mu K^{ij}{}_\nu - K^i{}_{m\mu}K^{mj}{}_\nu - (\mu \leftrightarrow \nu)\right].$$

This relation, combined with the field equations (2.5), leads to

$$\tilde{R}^{ij}{}_{\mu\nu} = -\Lambda_{\text{eff}}(b^i{}_\mu b^j{}_\nu - b^i{}_\nu b^j{}_\mu), \qquad \Lambda_{\text{eff}} \equiv q - \frac{1}{4}p^2, \tag{2.6}$$

where Λ_{eff} is the effective cosmological constant. Equation (2.6) can be considered as an equivalent of the second field equation (2.5b). Looking at (2.6) as an equation for the metric, one finds that our spacetime has maximally symmetric metric [25]:

- for $\Lambda_{\text{eff}} < 0$ ($\Lambda_{\text{eff}} > 0$), the spacetime manifold is anti-de Sitter (de Sitter).

There are two interesting special cases of the general Mielke-Baekler model, which have been studied in the past.

- For $\alpha_3 = \alpha_4 = 0$, the vacuum geometry becomes *Riemannian*, $T_{ijk} = 0$. This choice corresponds to GR$_\Lambda$ [3, 4];
- for $(\alpha_4)^2 + a\Lambda = 0$, the vacuum geometry is *teleparallel*, $R_{ijk} = 0$. The vacuum field equations are "geometrically dual" to those of GR$_\Lambda$ [20].

In the present paper, we shall investigate the general model (2.4) with $\alpha_3\alpha_4 - a^2 \neq 0$, assuming that the effective cosmological constant is negative (anti-de Sitter sector):

$$\Lambda_{\text{eff}} \equiv -\frac{1}{\ell^2} < 0. \tag{2.7}$$

The de Sitter sector with $\Lambda_{\text{eff}} > 0$ is left for the future studies.

3 Exact vacuum solutions

Some aspects of the canonical analysis rely on the existence of suitable asymptotic conditions. A proper choice of these conditions is based, to some extent, on the properties of exact classical solutions in vacuum. For the Mielke-Baekler model (2.4), these solutions are well known [19, 22]. Their construction can be described by the following set of rules:

- For a given Λ_{eff}, use Eq. (2.6) to find a solution for the metric. This step is very simple, since the metric structure of maximally symmetric spaces is well known [25].

- Given the metric, find a solution for the triad field, such that $g = \eta_{ij} b^i \otimes b^j$.
- Finally, use Eq. (2.5a) to determine the connection ω^i.

For exact solutions with non-vanishing sources, the reader can consult Ref. [24].

Riemann-Cartan black hole. For $\Lambda_{\text{eff}} < 0$, equation (2.6) has a well known solution for the metric — the BTZ black hole [5]. Using the static coordinates $x^\mu = (t, r, \varphi)$ (with $0 \leq \varphi < 2\pi$), and units $4G = 1$, it is given as

$$ds^2 = N^2 dt^2 - N^{-2} dr^2 - r^2(d\varphi + N_\varphi dt)^2,$$
$$N^2 = \left(-2m + \frac{r^2}{\ell^2} + \frac{J^2}{r^2}\right), \qquad N_\varphi = \frac{J}{r^2}. \tag{3.1}$$

The parameters m and J are related to the conserved charges—energy and angular momentum. Since the triad field corresponding to (3.1) is determined only up to a local Lorentz transformation, we can choose b^i to have the simple, "diagonal" form:

$$b^0 = N dt, \qquad b^1 = N^{-1} dr, \qquad b^2 = r(d\varphi + N_\varphi dt). \tag{3.2a}$$

Then, the connection is obtained by solving the first field equation (2.5a):

$$\omega^0 = N\left(\frac{p}{2}dt - d\varphi\right), \quad \omega^1 = N^{-1}\left(\frac{p}{2} + \frac{J}{r^2}\right)dr,$$
$$\omega^2 = -\left(\frac{r}{\ell} - \frac{p\ell}{2}\frac{J}{r}\right)\frac{dt}{\ell} + \left(\frac{p}{2}r - \frac{J}{r}\right)d\varphi. \tag{3.2b}$$

Equations (3.2) define the *Riemann-Cartan* black hole.

Riemann-Cartan AdS solution. In Riemannian geometry with negative Λ, the general solution with maximal number of Killing vectors is called the AdS solution [5, 25]. Although AdS solution and the black hole are locally isometric, they are globally distinct. The AdS solution is obtained from (3.1) by the substitution $J = 0, 2m = -1$.

Similarly, there is a general solution with maximal symmetry in Riemann-Cartan geometry, the *Riemann-Cartan* AdS solution. It is obtained from the black hole (3.2) by the same substitution ($J = 0, 2m = -1$). Using the notation $f^2 \equiv 1 + r^2/\ell^2$, we have:

$$b^0 = f dt, \quad b^1 = f^{-1} dr, \quad b^2 = r d\varphi, \tag{3.3a}$$
$$\omega^0 = f\left(\frac{p}{2}dt - d\varphi\right), \quad \omega^1 = \frac{p}{2f}dr, \quad \omega^2 = -\frac{r}{\ell}\left(\frac{dt}{\ell} - \frac{p\ell}{2}d\varphi\right). \tag{3.3b}$$

In order to understand symmetry properties of (3.3), we note that the form-invariance of a given field configuration in Riemann-Cartan geometry is defined by the requirements $\delta_0 b^i{}_\mu = 0$, $\delta_0 \omega^i{}_\mu = 0$, which differ from the Killing equation in Riemannian geometry, $\delta_0 g_{\mu\nu} = 0$ (δ_0 is the PGT analogue of the geometric Lie derivative). When applied to the Riemann-Cartan AdS solution (3.3), these requirements restrict (ξ^μ, θ^i) to the subspace defined by the basis of six pairs $(\xi^\mu_{(k)}, \theta^i_{(k)})$ ($k = 1, \ldots, 6$), given in Appendix A. The related symmetry group is the six-dimensional AdS group $SO(2,2)$.

4 Asymptotic conditions

Spacetime outside localized matter sources is described by the vacuum solutions of the field equations (2.5). Thus, matter has no influence on the local properties of spacetime in the source-free regions, but it can change its global properties. On the other hand, global properties of spacetime affect symmetry properties of the asymptotic configurations, and consequently, they are closely related to the gravitational conservation laws.

In 3D gravity with $\Lambda_{\text{eff}} < 0$, maximally symmetric AdS solution (3.3) has the role analogous to the role of Minkowski space in the $\Lambda_{\text{eff}} = 0$ case. Following this analogy, we could choose (3.3) to be the field configuration to which all the dynamical variables approach in such a way, that the asymptotic symmetry is $SO(2, 2)$, the maximal symmetry of (3.3). However, such an assumption would exclude the important black hole geometries, which are not $SO(2,2)$ invariant. Having an idea to maximally relax the asymptotic conditions and enlarge the set of asymptotic states (and the relevant group of symmetries), we introduce the concept of the *AdS asymptotic behavior*, based on the following requirements [3, 26]:

(a) asymptotic configurations should include the black hole geometries;

(b) they should be invariant under the action of the AdS group $SO(2,2)$;
(c) asymptotic symmetries should have well defined canonical generators.

The conditions (a) and (b) together lead to an extended asymptotic structure, quite different from the standard, form-invariant vacuum configuration, while (c) is just a technical assumption.

AdS asymptotics. We begin our considerations with the point (a) above. The asymptotic behaviour of the black hole triad (3.2a) is given by

$$b^i{}_\mu \sim \begin{pmatrix} \frac{r}{\ell} - \frac{m\ell}{r} & 0 & 0 \\ 0 & \frac{\ell}{r} + \frac{m\ell^3}{r^3} & 0 \\ \frac{J}{r} & 0 & r \end{pmatrix},$$

where the type of higher order terms on the right hand side is not written explicitly. Similarly, the asymptotic behaviour of the connection (3.2b) has the form

$$\omega^i{}_\mu \sim \begin{pmatrix} \frac{p\ell}{2}\left(\frac{r}{\ell^2} - \frac{m}{r}\right) & 0 & -\frac{r}{\ell} + \frac{m\ell}{r} \\ 0 & \frac{p\ell}{2r} + \frac{J\ell + pm\ell^3/2}{r^3} & 0 \\ -\frac{r}{\ell^2} + \frac{pJ}{2r} & 0 & \frac{pr}{2} - \frac{J}{r} \end{pmatrix}.$$

According to (a), asymptotic conditions should be chosen so as to *include* these black hole configurations.

In order to realize the requirement (b), we start with the above black hole configuration and act on it with all possible $SO(2,2)$ transformations, defined by the basis of six pairs $(\xi_{(k)}, \theta_{(k)})$, displayed in Appendix A. The result has the form

$$\delta_{(k)} b^i{}_\mu \sim \begin{pmatrix} \mathcal{O}_1 & \mathcal{O}_4 & \mathcal{O}_1 \\ \mathcal{O}_2 & \mathcal{O}_3 & \mathcal{O}_2 \\ \mathcal{O}_1 & \mathcal{O}_4 & \mathcal{O}_1 \end{pmatrix}, \qquad \delta_{(k)} \omega^i{}_\mu \sim \begin{pmatrix} \mathcal{O}_1 & \mathcal{O}_4 & \mathcal{O}_1 \\ \mathcal{O}_2 & \mathcal{O}_3 & \mathcal{O}_2 \\ \mathcal{O}_1 & \mathcal{O}_4 & \mathcal{O}_1 \end{pmatrix},$$

where $\mathcal{O}_n$ denotes a quantity that tends to zero as $1/r^n$ or faster, when $r \to \infty$.

The family of the black hole triads obtained in this way is parametrized by six real parameters, say σ_i; we denote it by $\mathcal{B}_6$. In order to have a set of asymptotic states which is sufficiently large to *include* the whole $\mathcal{B}_6$, we adopt the following asymptotic form for the triad field:

$$b^i{}_\mu = \begin{pmatrix} \frac{r}{\ell} + \mathcal{O}_1 & \mathcal{O}_4 & \mathcal{O}_1 \\ \mathcal{O}_2 & \frac{\ell}{r} + \mathcal{O}_3 & \mathcal{O}_2 \\ \mathcal{O}_1 & \mathcal{O}_4 & r + \mathcal{O}_1 \end{pmatrix} \equiv \begin{pmatrix} \frac{r}{\ell} & 0 & 0 \\ 0 & \frac{\ell}{r} & 0 \\ 0 & 0 & r \end{pmatrix} + B^i{}_\mu. \tag{4.1a}$$

The real meaning of this expression and its relation to $\mathcal{B}_6$ is clarified by noting that any c/r^n term in $\mathcal{B}_6$ is transformed into the corresponding $c(t,\varphi)/r^n$ term in (4.1a), i.e. constants $c = c(\sigma_i)$ are promoted to functions $c(t,\varphi)$. Thus, (4.1a) is a natural generalization of $\mathcal{B}_6$.

The triad family (4.1a) generates the Brown–Henneaux asymptotic form of the metric,

$$g_{\mu\nu} = \begin{pmatrix} \frac{r^2}{\ell^2}+\mathcal{O}_0 & \mathcal{O}_3 & \mathcal{O}_0 \\ \mathcal{O}_3 & -\frac{\ell^2}{r^2}+\mathcal{O}_4 & \mathcal{O}_3 \\ \mathcal{O}_0 & \mathcal{O}_3 & -r^2+\mathcal{O}_0 \end{pmatrix} \equiv \begin{pmatrix} \frac{r^2}{\ell^2} & 0 & 0 \\ 0 & -\frac{\ell^2}{r^2} & 0 \\ 0 & 0 & -r^2 \end{pmatrix} + G_{\mu\nu},$$

but clearly, it is not uniquely determined by it (any Lorentz transform of the triad produces the same metric).

Having found the triad asymptotics, we now use similar arguments to find the needed asymptotic behavior for the connection:

$$\omega^i{}_\mu = \begin{pmatrix} \frac{pr}{2\ell}+\mathcal{O}_1 & \mathcal{O}_2 & -\frac{r}{\ell}+\mathcal{O}_1 \\ \mathcal{O}_2 & \frac{p\ell}{2r}+\mathcal{O}_3 & \mathcal{O}_2 \\ -\frac{r}{\ell^2}+\mathcal{O}_1 & \mathcal{O}_2 & \frac{pr}{2}+\mathcal{O}_1 \end{pmatrix} \equiv \begin{pmatrix} \frac{pr}{2\ell} & 0 & -\frac{r}{\ell} \\ 0 & \frac{p\ell}{2r} & 0 \\ -\frac{r}{\ell^2} & 0 & \frac{pr}{2} \end{pmatrix} + \Omega^i{}_\mu. \quad (4.1b)$$

Note that the choice $\omega^0{}_1, \omega^2{}_1 = \mathcal{O}_2$, adopted in (4.1b), represents an acceptable generalization of the conditions $\omega^0{}_1, \omega^2{}_1 = \mathcal{O}_4$, suggested by the form of $\delta_{(k)}\omega^i{}_\mu$ (compare also with the conditions (C.1)).

As we have seen, the requirements (a) and (b) are not sufficient for a unique determination of the asymptotic behavior. Our choice of the asymptotics was guided by the idea to obtain the *most general* asymptotic behavior compatible with (a) and (b), with arbitrary higher-order terms $B^i{}_\mu$ and $\Omega^i{}_\mu$. Although $B^i{}_\mu$ and $\Omega^i{}_\mu$ are arbitrary at this stage, certain relations among them will be established latter (Appendix C), using some additional requirements. One can verify that the asymptotic conditions (4.1) are indeed invariant under the action of the AdS group $SO(2,2)$. In the next step, we shall examine whether there is any *higher* symmetry structure in (4.1), which will be the real test of our choice.

Asymptotic symmetries. Having chosen the asymptotic conditions in the form (4.1), we now wish to find the subset of gauge transformations that respect these conditions. Acting on a specific field satisfying (4.1), these transformations are allowed to change the form of the non-leading terms $B^i{}_\mu, \Omega^i{}_\mu$, as they are arbitrary by assumption. Thus, the parameters of the restricted gauge transformations are determined by the relations

$$\varepsilon^{ijk}\theta_j b_{k\mu} - (\partial_\mu \xi^\rho)b^i{}_\rho - \xi^\rho \partial_\rho b^i{}_\mu = \delta_0 B^i{}_\mu, \quad (4.2a)$$

$$-\partial_\mu \theta^i + \varepsilon^{ijk}\theta_j \omega_{k\mu} - (\partial_\mu \xi^\rho)\omega^i{}_\rho - \xi^\rho \partial_\rho \omega^i{}_\mu = \delta_0 \Omega^i{}_\mu. \quad (4.2b)$$

The transformations defined in this way differ from those that are associated to the form-invariant vacuum configurations ($\delta_0 b^i{}_\mu = 0, \delta_0 \omega^i{}_\mu = 0$). The restricted gauge parameters are determined as follows [20].

The symmetric part of (4.2a) multiplied by $b_{i\nu}$ (six relations) yields the transformation rule of the metric:

$$-(\partial_\mu \xi^\rho)g_{\nu\rho} - (\partial_\nu \xi^\rho)g_{\mu\rho} - \xi^\rho \partial_\rho g_{\mu\nu} = \delta_0 G_{\mu\nu}.$$

Expanding ξ^μ in powers of r^{-1}, we find the solution of these equations as

$$\xi^0 = \ell \left[T + \frac{1}{2}\left(\frac{\partial^2 T}{\partial t^2}\right)\frac{\ell^4}{r^2}\right] + \mathcal{O}_4\,, \tag{4.3a}$$

$$\xi^2 = S - \frac{1}{2}\left(\frac{\partial^2 S}{\partial\varphi^2}\right)\frac{\ell^2}{r^2} + \mathcal{O}_4\,, \tag{4.3b}$$

$$\xi^1 = -\ell\left(\frac{\partial T}{\partial t}\right) r + \mathcal{O}_1\,, \tag{4.3c}$$

where the functions $T(t,\varphi)$ and $S(t,\varphi)$ satisfy the conditions

$$\frac{\partial T}{\partial \varphi} = \ell\frac{\partial S}{\partial t}\,, \qquad \frac{\partial S}{\partial \varphi} = \ell\frac{\partial T}{\partial t}\,. \tag{4.4}$$

In GR_Λ, these equations define the two-dimensional conformal group at large distances [3].
The remaining three components of (4.2a) determine θ^i:

$$\theta^0 = -\frac{\ell^2}{r}\partial_0\partial_2 T + \mathcal{O}_3\,,$$

$$\theta^2 = \frac{\ell^3}{r}\partial_0^2 T + \mathcal{O}_3\,,$$

$$\theta^1 = \partial_2 T + \mathcal{O}_2\,. \tag{4.3d}$$

The conditions (4.2b) produce no new limitations on the parameters.

Introducing the light-cone coordinates $x^\pm = x^0/\ell \pm x^2$, the conditions (4.4) can be written in the form

$$\partial_\pm(T \mp S) = 0\,,$$

from which one easily finds the general solution for T and S:

$$T + S = g(x^+)\,, \qquad T - S = h(x^-)\,, \tag{4.5}$$

where g and h are two arbitrary, periodic functions.

The commutator algebra of Poincaré gauge transformations (2.1) is closed: we have $[\delta_0', \delta_0''] = \delta_0'''$, where $\delta_0' \equiv \delta_0(\xi',\theta')$ and so on, and the composition law reads:

$$\xi'''^\mu = \xi'^\rho\partial_\rho\xi''^\mu - \xi''^\rho\partial_\rho\xi'^\mu\,,$$

$$\theta'''^i = \varepsilon^i{}_{mn}\theta'^m\theta''^n + \xi'\cdot\partial\theta''^i - \xi''\cdot\partial\theta'^i\,.$$

Substituting here the restricted form of the parameters (4.3) and comparing the lowest order terms, we find the relations

$$T''' = T'\partial_2 S'' + S'\partial_2 T'' - T''\partial_2 S' - S''\partial_2 T'\,,$$
$$S''' = S'\partial_2 S'' + T'\partial_2 T'' - S''\partial_2 S' - T''\partial_2 T'\,, \tag{4.6}$$

that are expected to be the composition law for (T, S). To clarify the situation, consider the restricted form of the gauge parameters (4.3), and separate it into two pieces: the leading terms containing T and S, which define the (T, S) transformations, and the higher order terms that remain after imposing $T = S = 0$, which define the *residual* (or pure) gauge transformations. If the relations (4.6) are to represent the composition law for the (T, S) transformations, one has to check their consistency with higher order terms in the commutator algebra. As one can verify, the commutator of two (T, S) transformations produces not only a (T, S) transformation, with the composition law (4.6), but also an additional, pure gauge transformation. However, pure gauge transformations are irrelevant for our discussion of the conservation laws. Indeed, as we shall see in section 6, they do not contribute to the values of the conserved charges (their generators vanish weakly). Thus, we are naturally led to correct the non-closure of the (T, S) commutator algebra by introducing an improved definition of the *asymptotic symmetry* [3, 26]:

- the asymptotic symmetry group is defined as the factor group of the gauge group determined by (4.3), with respect to the residual gauge group.

In other words, two asymptotic transformations are identified if they have the same (T, S) pairs. The asymptotic symmetry of our spacetime represents an unusual realization of the *conformal symmetry* (see section 6).

The set of asymptotic conditions (4.1) is invariant under the group of transformations with parameters (4.3), which is much larger then the original AdS group $SO(2, 2)$. The resulting configuration space respects the requirements (a) and (b) formulated at the beginning of this section. The asymptotic structure of the whole phase space, as well as the status of the last requirement (c), will be examined in the next two sections.

5 Gauge generator

In gauge theories, the presence of unphysical variables is closely related to the existence of gauge symmetries. The best way to understand the dynamical content of these symmetries is to explore the related canonical generator, which acts on the basic dynamical variables via the PB operation. To begin the analysis, we rewrite the action (2.4) as

$$I = \int d^3x \varepsilon^{\mu\nu\rho} \left[ab^i{}_\mu R_{i\nu\rho} - \frac{1}{3}\Lambda \varepsilon_{ijk} b^i{}_\mu b^j{}_\nu b^k{}_\rho \right.$$
$$\left. + \alpha_3 \left(\omega^i{}_\mu \partial_\nu \omega_{i\rho} + \frac{1}{3}\varepsilon_{ijk}\omega^i{}_\mu \omega^j{}_\nu \omega^k{}_\rho \right) + \frac{1}{2}\alpha_4 b^i{}_\mu T_{i\nu\rho} \right]. \quad (5.1)$$

Hamiltonian and constraints. The basic Lagrangian variables $(b^i{}_\mu, \omega^i{}_\mu)$ and the corresponding canonical momenta $(\pi^i{}_\mu, \Pi^i{}_\mu)$ are related to each other through the set of primary constraints:

$$\phi_i{}^0 \equiv \pi_i{}^0 \approx 0, \qquad \Phi_i{}^0 \equiv \Pi_i{}^0 \approx 0,$$
$$\phi_i{}^\alpha \equiv \pi_i{}^\alpha - \alpha_4 \varepsilon^{0\alpha\beta} b_{i\beta} \approx 0, \qquad \Phi_i{}^\alpha \equiv \Pi_i{}^\alpha - \varepsilon^{0\alpha\beta}(2ab_{i\beta} + \alpha_3 \omega_{i\beta}) \approx 0. \quad (5.2)$$

Explicit construction of the canonical Hamiltonian yields an expression which is linear in unphysical variables, as expected:

$$\mathcal{H}_c = b^i{}_0 \mathcal{H}_i + \omega^i{}_0 \mathcal{K}_i + \partial_\alpha D^\alpha,$$

$$\mathcal{H}_i = -\varepsilon^{0\alpha\beta}\left(aR_{i\alpha\beta} + \alpha_4 T_{i\alpha\beta} - \Lambda\varepsilon_{ijk}b^j{}_\alpha b^k{}_\beta\right),$$

$$\mathcal{K}_i = -\varepsilon^{0\alpha\beta}\left(aT_{i\alpha\beta} + \alpha_3 R_{i\alpha\beta} + \alpha_4\varepsilon_{ijk}b^j{}_\alpha b^k{}_\beta\right),$$

$$D^\alpha = \varepsilon^{0\alpha\beta}\left[\omega^i{}_0\left(2ab_{i\beta} + \alpha_3\omega_{i\beta}\right) + \alpha_4 b^i{}_0 b_{i\beta}\right].$$

Going over to the total Hamiltonian,

$$\mathcal{H}_T = b^i{}_0\mathcal{H}_i + \omega^i{}_0\mathcal{K}_i + u^i{}_\mu\phi_i{}^\mu + v^i{}_\mu\Phi_i{}^\mu + \partial_\alpha D^\alpha, \tag{5.3}$$

we find that the consistency conditions of the sure primary constraints $\pi_i{}^0$ and $\Pi_i{}^0$ yield the secondary constraints:

$$\mathcal{H}_i \approx 0, \qquad \mathcal{K}_i \approx 0. \tag{5.4a}$$

These constraints can be equivalently written in the form:

$$T_{i\alpha\beta} \approx p\varepsilon_{ijk}b^j{}_\alpha b^k{}_\beta, \qquad R_{i\alpha\beta} \approx q\varepsilon_{ijk}b^j{}_\alpha b^k{}_\beta. \tag{5.4b}$$

The consistency of the remaining primary constraints $\phi^i{}_\alpha$ and $\Phi^i{}_\alpha$ leads to the determination of the multipliers $u^i{}_\beta$ and $v^i{}_\beta$ (see Appendix B):

$$\begin{aligned}u^i{}_\beta + \varepsilon^{ijk}\omega_{j0}b_{k\beta} - \nabla_\beta b^i_0 &= p\varepsilon^{ijk}b_{j0}b_{k\beta},\\ v^i{}_\beta - \nabla_\beta \omega^i{}_0 &= q\varepsilon^{ijk}b_{j0}b_{k\beta}.\end{aligned} \tag{5.5a}$$

Using the equations of motion $\dot{b}^i{}_\beta = u^i{}_\beta$ and $\dot{\omega}^i{}_\beta = v^i{}_\beta$, these relations reduce to the field equations

$$T^i{}_{0\beta} \approx p\varepsilon^{ijk}b_{j0}b_{k\beta}, \qquad R^i{}_{0\beta} \approx q\varepsilon^{ijk}b_{j0}b_{k\beta}. \tag{5.5b}$$

The substitution of the determined multipliers (5.5a) into (5.3) yields the final form of the total Hamiltonian:

$$\begin{aligned}\mathcal{H}_T &= \hat{\mathcal{H}}_T + \partial_\alpha \bar{D}^\alpha,\\ \hat{\mathcal{H}}_T &= b^i{}_0\bar{\mathcal{H}}_i + \omega^i{}_0\bar{\mathcal{K}}_i + u^i{}_0\pi_i{}^0 + v^i{}_0\Pi_i{}^0,\end{aligned} \tag{5.6a}$$

where

$$\begin{aligned}\bar{\mathcal{H}}_i &= \mathcal{H}_i - \nabla_\beta\phi_i{}^\beta + \varepsilon_{ijk}b^j{}_\beta\left(p\phi^{k\beta} + q\Phi^{k\beta}\right),\\ \bar{\mathcal{K}}_i &= \mathcal{K}_i - \nabla_\beta\Phi_i{}^\beta - \varepsilon_{ijk}b^j{}_\beta\phi^{k\beta},\\ \bar{D}^\alpha &= D^\alpha + b^i{}_0\phi_i{}^\alpha + \omega^i{}_0\Phi_i{}^\alpha.\end{aligned} \tag{5.6b}$$

Further investigation of the consistency procedure is facilitated by observing that the secondary constraints $\bar{\mathcal{H}}_i, \bar{\mathcal{K}}_i$ obey the PB relations (B.2). One concludes that the consistency conditions of the secondary constraints (5.4) are identically satisfied, which completes the Hamiltonian consistency procedure.

Complete classification of the constraints is given in the following table.

Table 1. Classification of constraints

	First class	Second class
Primary	$\pi_i{}^0, \Pi_i{}^0$	$\phi_i{}^\alpha, \Phi_i{}^\alpha$
Secondary	$\bar{\mathcal{H}}_i, \bar{\mathcal{K}}_i$	

Canonical gauge generator. The results of the previous analysis are sufficient for the construction of the gauge generator [27]. Starting from the primary first class constraints $\pi_i{}^0$ and $\Pi_i{}^0$, one obtaines:

$$\begin{aligned} G[\epsilon] &= \dot{\epsilon}^i \pi_i{}^0 + \epsilon^i \left[\bar{\mathcal{H}}_i - \varepsilon_{ijk} \left(\omega^j{}_0 - p b^j{}_0 \right) \pi^{k0} + q \varepsilon_{ijk} b^j{}_0 \Pi^{k0} \right], \\ G[\tau] &= \dot{\tau}^i \Pi_i{}^0 + \tau^i \left[\bar{\mathcal{K}}_i - \varepsilon_{ijk} \left(b^j{}_0 \pi^{k0} + \omega^j{}_0 \Pi^{k0} \right) \right]. \end{aligned} \qquad (5.7)$$

The complete gauge generator is given by the expression $G = G[\epsilon] + G[\tau]$, and its action on the fields, defined by $\delta_0 \phi = \{\phi, G\}$, has the form:

$$\begin{aligned} \delta_0 b^i{}_\mu &= \nabla_\mu \epsilon^i - p \varepsilon^i{}_{jk} b^j{}_\mu \tau^k + \varepsilon^i{}_{jk} b^j{}_\mu \tau^k, \\ \delta_0 \omega^i{}_\mu &= \nabla_\mu \tau^i - q \varepsilon^i{}_{jk} b^j{}_\mu \epsilon^k. \end{aligned}$$

This result looks more like a standard gauge transformation, with no trace of the expected local Poincaré transformations. However, after introducing the new parameters

$$\epsilon^i = -\xi^\mu b^i{}_\mu, \qquad \tau^i = -(\theta^i + \xi^\mu \omega^i{}_\mu),$$

one easily obtains

$$\begin{aligned} \delta_0 b^i{}_\mu &= \delta_{\text{PGT}} b^i{}_\mu - \xi^\rho \left(T^i{}_{\mu\rho} - p \varepsilon^{ijk} b_{j\mu} b_{k\rho} \right), \\ \delta_0 \omega^i{}_\mu &= \delta_{\text{PGT}} \omega^i{}_\mu - \xi^\rho \left(R^i{}_{\mu\rho} - q \varepsilon^{ijk} b_{j\mu} b_{k\rho} \right). \end{aligned}$$

Thus, *on-shell*, we have the transformation laws that are in complete agreement with (2.1). Expressed in terms of the new parameters, the gauge generator takes the form

$$\begin{aligned} G &= -G_1 - G_2, \\ G_1 &\equiv \dot{\xi}^\rho \left(b^i{}_\rho \pi_i{}^0 + \omega^i{}_\rho \Pi_i{}^0 \right) + \xi^\rho \left[b^i{}_\rho \bar{\mathcal{H}}_i + \omega^i{}_\rho \bar{\mathcal{K}}_i + (\partial_\rho b^i_0) \pi_i{}^0 + (\partial_\rho \omega^i{}_0) \Pi^i{}_0 \right], \\ G_2 &\equiv \dot{\theta}^i \Pi_i{}^0 + \theta^i \left[\bar{\mathcal{K}}_i - \varepsilon_{ijk} \left(b^j{}_0 \pi^{k0} + \omega^j{}_0 \Pi^{k0} \right) \right], \end{aligned} \qquad (5.8)$$

where the time derivatives $\dot{b}^i{}_\mu$ and $\dot{\omega}^i{}_\mu$ are shorts for $u^i{}_\mu$ and $v^i{}_\mu$, respectively. Note, in particular, that the time translation generator is determined by the total Hamiltonian:

$$G\left[\xi^0\right] = -\dot{\xi}^0 \left(b^i{}_0 \pi_i{}^0 + \omega^i{}_0 \Pi_i{}^0 \right) - \xi^0 \hat{\mathcal{H}}_T.$$

In the above expressions, the integration symbol $\int d^3 x$ is omitted for simplicity; later, when necessary, it will be restored.

Asymptotics of the phase space. In order to extend the asymptotic conditions (4.1) to the canonical level, one should determine an appropriate asymptotic behavior of the whole phase space, including the momentum variables. This step is based on the following general principle:

- the expressions than vanish on shell should have an arbitrary fast asymptotic decrease, as no solutions of the field equations are thereby lost.

By applying this principle to the primary constraints (5.2), one finds the following asymptotic behavior of the momentum variables:

$$\pi_i{}^0 = \hat{\mathcal{O}}, \qquad \pi_i{}^\alpha = \alpha_4 \varepsilon^{0\alpha\beta} b_{i\beta} + \hat{\mathcal{O}}$$
$$\Pi_i{}^0 = \hat{\mathcal{O}}, \qquad \Pi_i{}^\alpha = \varepsilon^{0\alpha\beta}(2ab_{i\beta} + \alpha_3 \omega_{i\beta}) + \hat{\mathcal{O}}. \qquad (5.9)$$

We shall use this principle again in connection to the consistency requiremets (5.4b) and (5.5b), in order to *refine* the general asymptotic canditions (4.1) and (5.9) (Appendix C).

6 Canonical realization of the asymptotic symmetry

In this section, we study the influence of the adopted asymptotics on the canonical structure of the theory: we construct the improved gauge generators, examine their canonical algebra and prove the conservation laws.

Improving the generators. The canonical generator acts on dynamical variables via the PB operation, which is defined in terms of functional derivatives. A phase-space functional $F = \int d^2 x f(\phi, \partial\phi, \pi, \partial\pi)$ has well defined functional derivatives if its variation can be written in the form $\delta F = \int d^2x \, [A(x)\delta\phi(x) + B(x)\delta\pi(x)]$, where terms $\delta\phi_{,\mu}$ and $\delta\pi_{,\nu}$ are absent. In order to ensure this property for our generator (5.8), we have to improve its form by adding certain surface terms [28].

Let us start the procedure by examining the variations of G_2:

$$\delta G_2 = \theta^i \delta \bar{\mathcal{K}}_i + R = \theta^i \delta \mathcal{K}_i + \partial\hat{\mathcal{O}} + R$$
$$= -2\varepsilon^{0\alpha\beta} \theta^i \left(a \partial_\alpha \delta b_{i\beta} + \alpha_3 \partial_\alpha \delta \omega_{i\beta} \right) + \partial\hat{\mathcal{O}} + R$$
$$= -2\varepsilon^{0\alpha\beta} \partial_\alpha \left(a \theta^i \delta b_{i\beta} + \alpha_3 \theta^i \delta \omega_{i\beta} \right) + \partial\hat{\mathcal{O}} + R = \partial\mathcal{O}_2 + R,$$

where the last equality follows from the asymptotic relations $\theta^i \delta b_{i\beta}, \theta^i \delta \omega_{i\beta} = \mathcal{O}_2$. The total divergence term $\partial\mathcal{O}_2$ gives a vanishing contribution after integration, as follows from the Stokes theorem:

$$\int_{\mathcal{M}_2} d^2 x \, \partial_\alpha v^\alpha = \int_{\partial\mathcal{M}_2} v^\alpha df_\alpha = \int_0^{2\pi} v^1 d\varphi \qquad (df_\alpha = \varepsilon_{\alpha\beta} dx^\beta),$$

where the boundary of the spatial section $\mathcal{M}_2$ of spacetime is taken to be the circle at infinity, parametrized by the angular coordinate φ. Thus, the boundary term for G_2 vanishes, and G_2 is regular as it stands, without any correction.

Going over to G_1, we have:

$$\delta G_1 = \xi^\rho \left(b^i{}_\rho \delta \bar{\mathcal{H}}_i + \omega^i{}_\rho \delta \bar{\mathcal{K}}_i \right) + \partial\hat{\mathcal{O}} + R$$
$$= -2\varepsilon^{0\alpha\beta} \partial_\alpha \left[\xi^\rho b^i{}_\rho \delta (a\omega_{i\beta} + \alpha_4 b_{i\beta}) + \xi^\rho \omega^i{}_\rho \delta (ab_{i\beta} + \alpha_3 \omega_{i\beta}) \right] + \partial\hat{\mathcal{O}} + R.$$

Using the adopted asymptotic conditions, the preceding result leads to

$$\delta G_1 = -\partial_\alpha \left(\xi^0 \delta \mathcal{E}^\alpha + \xi^2 \delta \mathcal{M}^\alpha \right) + \partial\mathcal{O}_2 + R$$
$$= -\delta \partial_\alpha \left(\xi^0 \mathcal{E}^\alpha + \xi^2 \mathcal{M}^\alpha \right) + \partial\mathcal{O}_2 + R,$$

where

$$\mathcal{E}^\alpha \equiv 2\varepsilon^{0\alpha\beta}\left[\left(a + \frac{\alpha_3 p}{2}\right)\omega^0{}_\beta + \left(\alpha_4 + \frac{ap}{2}\right)b^0{}_\beta + \frac{a}{\ell}b^2{}_\beta + \frac{\alpha_3}{\ell}\omega^2{}_\beta\right]b^0{}_0,$$

$$\mathcal{M}^\alpha \equiv -2\varepsilon^{0\alpha\beta}\left[\left(a + \frac{\alpha_3 p}{2}\right)\omega^2{}_\beta + \left(\alpha_4 + \frac{ap}{2}\right)b^2{}_\beta + \frac{a}{\ell}b^0{}_\beta + \frac{\alpha_3}{\ell}\omega^0{}_\beta\right]b^2{}_2. \quad (6.1)$$

Thus, the improved form of the complete gauge generator (5.8) reads:

$$\tilde{G} = G + \Gamma,$$
$$\Gamma = -\oint df_\alpha\left(\xi^0 \mathcal{E}^\alpha + \xi^2 \mathcal{M}^\alpha\right) = -\int_0^{2\pi} d\varphi\left(\ell T \mathcal{E}^1 + S \mathcal{M}^1\right). \quad (6.2)$$

The adopted asymptotic conditions guarantee that $\tilde{G}$ is finite and differentiable functional. The boundary term Γ depends only on T and S, not on any pure gauge term in (4.3).

The improved time translation generator has the form

$$\tilde{G}[\xi^0] = G[\xi^0] - E[\xi^0], \qquad E[\xi^0] \equiv \int_0^{2\pi} d\varphi\,\xi^0 \mathcal{E}^1. \quad (6.3a)$$

For $\xi^0 = 1$, the generator G reduces to $-\hat{H}_T$, and the corresponding boundary term has the meaning of energy:

$$\tilde{H}_T = \hat{H}_T + E, \qquad E = \int_0^{2\pi} d\varphi\,\mathcal{E}^1. \quad (6.3b)$$

The improved spatial rotation generator is given by

$$\tilde{G}[\xi^2] = G[\xi^2] - M[\xi^2], \qquad M[\xi^2] \equiv \int_0^{2\pi} d\varphi\,\xi^2 \mathcal{M}^1, \quad (6.4a)$$

where M is a finite integral. The boundary term for $\xi^2 = 1$ is the angular momentum of the system:

$$M = \int_0^{2\pi} d\varphi\,\mathcal{M}^1. \quad (6.4b)$$

Canonical algebra. The PB algebra of the improved generators could be found by a direct calculation, but we shall rather use another, more instructive method, based on the results of Refs. [29] and [20]. Let us first recall that our improved generator (6.2) is a differentiable phase space functional that preserves the asymptotic conditions (4.1) and (5.9), hence, it satisfies the conditions of the main theorem in Ref. [29]. Introducing a convenient notation, $\tilde{G}' \equiv \tilde{G}[T', S']$, $\tilde{G}'' \equiv \tilde{G}[T'', S'']$, the main theorem states that the PB $\{\tilde{G}'', \tilde{G}'\}$ of two differentiable generators is itself a differentiable generator. Taking into account that any differentiable generator $\tilde{G}$ is defined only up to an additive, constant phase-space functional C (which does not change the action of $\tilde{G}$ on the phase space), the main theorem leads directly to

$$\{\tilde{G}'', \tilde{G}'\} = \tilde{G}''' + C''', \quad (6.5)$$

where the parameters of $\tilde{G}'''$ are defined by the composition law (4.6), and C''' is an unknown, field-independent functional, $C''' \equiv C'''[T', S'; T'', S'']$. The term C''' is known as the *central charge* of the PB algebra.

In order to calculate C''', we note that $\{\tilde{G}'', \tilde{G}'\} = \delta_0' \tilde{G}'' \approx \delta_0' \Gamma''$, where the weak equality follows from the fact that δ_0' is a symmetry operation that maps constraints into constraints. Combining this result with $\tilde{G}''' \approx \Gamma'''$, Eq. (6.5) implies

$$\delta_0' \Gamma'' \approx \Gamma''' + C'''. \tag{6.6a}$$

This relation determines the value of C''' only weakly, but since C''' is a field-independent quantity, the weak equality is easily converted into the strong one. The calculation of $\delta_0' \Gamma''$ is based on the relations

$$\delta_0 \left(\ell \mathcal{E}^1 \right) = -\mathcal{M}^1 \partial_2 T - \ell \mathcal{E}^1 \partial_2 S - \partial_2 \left(\mathcal{M}^1 T + \ell \mathcal{E}^1 S \right)$$
$$+ (2a + \alpha_3 p) \ell \partial_2^3 S - 2\alpha_3 \partial_2^3 T + \mathcal{O}_2 \,,$$
$$\delta_0 \mathcal{M}^1 = -\ell \mathcal{E}^1 \partial_2 T - \mathcal{M}^1 \partial_2 S - \partial_2 \left(\ell \mathcal{E}^1 T + \mathcal{M}^1 S \right)$$
$$+ (2a + \alpha_3 p) \ell \partial_2^3 T - 2\alpha_3 \partial_2^3 S + \mathcal{O}_2 \,,$$

which are derived using the refined asymptotic conditions found in Appendix C, and the transformation rules defined by the parameters (4.3). The calculated expression for $\delta_0' \Gamma''$ has the form (6.6a), with the following value for the central charge C''':

$$C''' = (2a + \alpha_3 p) \ell \int_0^{2\pi} d\varphi \left(\partial_2 S'' \partial_2^2 T' - \partial_2 S' \partial_2^2 T'' \right)$$
$$- 2\alpha_3 \int_0^{2\pi} d\varphi \left(\partial_2 T'' \partial_2^2 T' + \partial_2 S'' \partial_2^2 S' \right) . \tag{6.6b}$$

Conservation laws. As we noted in section 5, the improved total Hamiltonian is one of the generators, $\tilde{G}[1, 0] = -\ell \tilde{H}_T$. A direct calculation based on the PB algebra (6.5) shows that the asymptotic generator $\tilde{G}[T, S]$ is conserved [20]:

$$\frac{d}{dt} \tilde{G} = \frac{\partial}{\partial t} \tilde{G} + \{\tilde{G}, \tilde{H}_T\} \approx \frac{\partial}{\partial t} \Gamma[T, S] - \frac{1}{\ell} \Gamma[\partial_2 S, \partial_2 T] = 0 \,. \tag{6.7}$$

This also implies the conservation of the boundary term Γ.

Now, we wish to clarify the meaning of the conserved charges by calculating their values for the black hole solution (3.2). First, note that the black hole solution depends on the radial coordinate only, and consequently, the terms $\mathcal{E}^1$ and $\mathcal{M}^1$ in Γ behave as constants. Second, the parameters (T, S) are periodic functions, equation (4.5), so that only *zero modes* in their Fourier expansion survive the integration in Γ. Thus, there are only two independent non-vanishing charges for the black hole solution, given by two inequivalent choices of the *constants* T and S. If we take, for instance, $(T = 1, S = 0)$ as the first choice, and $(T = 0, S = 1)$ as the second one, all the other non-zero charges will be given as linear combinations of these two.

For $(T = 1, S = 0)$ we have $\Gamma[1, 0] = -\ell E$, and the corresponding conserved charge is the energy E. Its value for the black hole solution is found to be

$$E(\text{black hole}) = m + \frac{\alpha_3}{a} \left(\frac{pm}{2} - \frac{J}{\ell^2} \right) . \tag{6.8a}$$

The second choice ($T = 0, S = 1$) leads to $\Gamma[0, 1] = -M$. The corresponding conserved charge is the angular momentum M, and its black hole value reads

$$M(\text{black hole}) = J + \frac{\alpha_3}{a}\left(\frac{pJ}{2} - m\right). \tag{6.8b}$$

Our expressions for the conserved charges (6.8) coincide with the results obtained in Ref. [19]. In the sector $\alpha_3 = 0$ (GR$_\Lambda$ and the teleparallel theory), we have $E = m$ and $M = J$, while for $\alpha_3 \neq 0$, the constants m and J do not have directly the meaning of energy and angular momentum, respectively. Geometrically, the two independent charges (6.8) parametrize the family of globally inequivalent, asymptotically AdS spaces.

Central charge. Expressed in terms of the Fourier modes, the canonical algebra (6.5) takes a more familiar form. Using the Fourier expansion for the parameters,

$$T = \sum_{-\infty}^{+\infty}\left(a_n e^{inx^+} + \bar{a}_n e^{inx^-}\right), \quad S = \sum_{-\infty}^{+\infty}\left(a_n e^{inx^+} - \bar{a}_n e^{inx^-}\right),$$

the canonical generator can be written in the form

$$\tilde{G}[T, S] = -2\sum_{-\infty}^{+\infty}\left(a_n L_n + \bar{a}_n \bar{L}_n\right), \tag{6.9a}$$

where

$$2L_n = -\tilde{G}[T = -S = e^{inx^-}], \quad 2\bar{L}_n = -\tilde{G}[T = S = e^{inx^+}].$$

Expressed in terms of the Fourier coefficients L_n and $\bar{L}_n$, the canonical algebra takes the form of two independent Virasoro algebras with classical central charges:

$$\{L_n, L_m\} = -i(n - m)L_{n+m} - \frac{c}{12}in^3\delta_{n,-m},$$
$$\{\bar{L}_n, \bar{L}_m\} = -i(n - m)\bar{L}_{m+n} - \frac{\bar{c}}{12}in^3\delta_{n,-m},$$
$$\{L_n, \bar{L}_m\} = 0. \tag{6.9b}$$

The central charges, in the standard string theory normalization, have the form:

$$c = \frac{3\ell}{2G} + 24\pi\alpha_3\left(\frac{p\ell}{2} + 1\right),$$
$$\bar{c} = \frac{3\ell}{2G} + 24\pi\alpha_3\left(\frac{p\ell}{2} - 1\right). \tag{6.10}$$

Thus, the gravitational sector with $\alpha_3 \neq 0$ has the conformal asymptotic symmetry with two different central charges, while $\alpha_3 = 0$ implies $c = \bar{c} = 3\ell/2G$.

- The general classical central charges c and $\bar{c}$ differ from each other, in contrast to the results obtained in GR$_\Lambda$ or the teleparallel theory [3, 20].

By redefining the zero modes, $L_0 \to L_0 + c/24$, $\bar{L}_0 \to \bar{L}_0 + \bar{c}/24$, the Virasoro algebra takes its standard form. One should note that the central term for the $SO(2, 2)$ subgroup, generated by $(L_\pm, L_0)$ and $(\bar{L}_\pm, \bar{L}_0)$, vanishes. This is a consequence of the fact that $SO(2, 2)$ is an exact symmetry of the AdS solution [3].

7 Concluding remarks

In this paper, we investigated the canonical structure of 3D gravity with torsion.

(1) The geometric arena for the topological 3D gravity with torsion, defined by the Mielke-Baekler action (2.4), has the form of Riemann-Cartan spacetime.

(2) There exists an exact vacuum solution of the theory, the Riemann-Cartan black hole (3.2), which generalizes the standard BTZ black hole in GR_Λ.

(3) Assuming the AdS asymptotic conditions, we constructed the canonical conserved charges. Energy and angular momentum of the Riemann-Cartan black hole are different from the corresponding BTZ values.

(4) The PB algebra of the canonical generators has the form of two independent Virasoro algebras with classical central charges. The values of the central charges are different from each other, in contrast to the situation in GR_Λ and the teleparallel theory. The implications of this result for the quantum structure of black hole are to be explored.

Acknowledgements

This work was supported by the Serbian Science foundation, Serbia. One of us (MB) would like to thank Milovan Vasilić, Friedrich Hehl and Yuri Obukhov for a critical reading of the manuscript and useful suggestions.

A Symmetries of the AdS vacuum

The invariance conditions $\delta_0 b^i{}_\mu = 0$ for the AdS triad (3.3a) yield the set of requirements on the parameters (ξ^μ, θ^i), the general solution of which has the following form [20]:

$$\xi^0 = \ell\sigma_1 - \frac{r}{f}\partial_2 Q, \qquad \xi^1 = \ell^2 f \partial_0 \partial_2 Q, \qquad \xi^2 = \sigma_2 - \frac{\ell^2 f}{r}\partial_0 Q,$$

$$\theta^0 = -\frac{\ell^2}{r}\partial_0 Q, \qquad \theta^1 = Q, \qquad \theta^2 = \frac{1}{f}\partial_2 Q, \qquad (A.1)$$

where

$$Q \equiv \sigma_3 \cos x^+ + \sigma_4 \sin x^+ + \sigma_5 \cos x^- + \sigma_6 \sin x^-, \qquad (A.2)$$

and σ_i are six arbitrary dimensionless parameters. The invariance conditions for the AdS connection (3.3b), $\delta_0 \omega^i{}_\mu = 0$, do not produce any new restrictions on (ξ^μ, θ^i). For each $k = 1, 2, \ldots, 6$, we can choose $\sigma_k = 1$ as the only non-vanishing constant, and find the corresponding basis of six independent Killing vectors $\xi^\mu_{(k)}$:

$$\xi_{(1)} = (\ell, 0, 0),$$
$$\xi_{(2)} = (0, 0, 1),$$
$$\xi_{(3)} = \left(\frac{r}{f}\sin x^+, -\ell f \cos x^+, \frac{\ell f}{r}\sin x^+\right),$$
$$\xi_{(4)} = \left(\frac{r}{f}\cos x^+, \ell f \sin x^+, \frac{\ell f}{r}\cos x^+\right),$$

$$\xi_{(5)} = \left(-\frac{r}{f}\sin x^-, \ell f \cos x^-, \frac{\ell f}{r}\sin x^-\right),$$
$$\xi_{(6)} = \left(\frac{r}{f}\cos x^-, \ell f \sin x^-, -\frac{\ell f}{r}\cos x^-\right), \qquad (A.3)$$

and similarly for $\theta^i_{(k)}$. As one can explicitly verify, the six pairs $(\xi^\mu_{(k)}, \theta^i_{(k)})$ fall into the class of asymptotic parameters (4.3), and define the algebra of the AdS group $SO(2,2)$.

B The algebra of constraints

The structure of the PB algebra of constraints is an important ingredient in the analysis of the Hamiltonian consistency conditions. For the nontrivial part of the PB algebra involving $(\phi_i{}^\alpha, \Phi_i{}^\alpha, \mathcal{H}_i, \mathcal{K}_i)$, we have the following result:

$$\begin{aligned}
\{\phi_i{}^\alpha, \phi_j{}^\beta\} &= -2\alpha_4 \varepsilon^{0\alpha\beta} \eta_{ij}\delta, \qquad \{\phi_i{}^\alpha, \Phi_j{}^\beta\} = -2a\varepsilon^{0\alpha\beta}\eta_{ij}\delta, \\
\{\Phi_i{}^\alpha, \Phi_j{}^\beta\} &= -2\alpha_3 \varepsilon^{0\alpha\beta}\eta_{ij}\delta, \\
\{\phi_i{}^\alpha, \mathcal{H}_j\} &= 2\varepsilon^{0\alpha\beta}\left[\alpha_4\eta_{ij}\partial_\beta\delta - \varepsilon_{ijk}\left(\alpha_4\omega^k{}_\beta - \Lambda b^k{}_\beta\right)\delta\right], \\
\{\phi_i{}^\alpha, \mathcal{K}_j\} &= 2\varepsilon^{0\alpha\beta}\left[a\eta_{ij}\partial_\beta\delta - \varepsilon_{ijk}\left(a\omega^k{}_\beta + \alpha_4 b^k{}_\beta\right)\delta\right], \\
\{\Phi_i{}^\alpha, \mathcal{H}_j\} &= 2\varepsilon^{0\alpha\beta}\left[a\eta_{ij}\partial_\beta\delta - \varepsilon_{ijk}\left(a\omega^k{}_\beta + \alpha_4 b^k{}_\beta\right)\delta\right], \\
\{\Phi_i{}^\alpha, \mathcal{K}_j\} &= 2\varepsilon^{0\alpha\beta}\left[\alpha_3\eta_{ij}\partial_\beta\delta - \varepsilon_{ijk}\left(\alpha_3\omega^k{}_\beta + a b^k{}_\beta\right)\delta\right]. \qquad (B.1)
\end{aligned}$$

The essential part of the PB algebra involving the first class constraints $(\bar{\mathcal{H}}_i, \bar{\mathcal{K}}_i)$ is given by the following relations:

$$\begin{aligned}
\{\phi_i{}^\alpha, \bar{\mathcal{H}}_j\} &= \varepsilon_{ijk}\left(p\phi^{k\alpha} + q\Phi^{k\alpha}\right)\delta, \qquad & \{\phi_i{}^\alpha, \bar{\mathcal{K}}_j\} &= -\varepsilon_{ijk}\phi^{k\alpha}\delta, \\
\{\Phi_i{}^\alpha, \bar{\mathcal{H}}_j\} &= -\varepsilon_{ijk}\phi^{k\alpha}\delta, \qquad & \{\Phi_i{}^\alpha, \bar{\mathcal{K}}_j\} &= -\varepsilon_{ijk}\Phi^{k\alpha}\delta, \\
\{\bar{\mathcal{H}}_i, \bar{\mathcal{H}}_j\} &= \varepsilon_{ijk}\left(p\bar{\mathcal{H}}^k + q\bar{\mathcal{K}}^k\right)\delta, \qquad & \{\bar{\mathcal{H}}_i, \bar{\mathcal{K}}_j\} &= -\varepsilon_{ijk}\bar{\mathcal{H}}^k\delta, \\
\{\bar{\mathcal{K}}_i, \bar{\mathcal{K}}_j\} &= -\varepsilon_{ijk}\bar{\mathcal{K}}^k\delta. & & \qquad (B.2)
\end{aligned}$$

C Asymptotic form of the constraints

Here, we analyze the influence of the secondary constraints (5.4) and relations (5.5) for the determined multipliers, on the basic asymptotic conditions (4.1) and (5.9), using the principle formulated at the end of section 5.

Let us start with the secondary constraints (5.4b). Using (4.1) and (5.9), these constraints imply the following asymptotic relations:

$$\begin{aligned}
\omega^0{}_1 &= \mathcal{O}_4, \qquad & \omega^2{}_1 &= \mathcal{O}_4, \\
\partial_1(re_2) &= \mathcal{O}_3, \qquad & \partial_1(rm_2) &= \mathcal{O}_3, \\
\partial_1\left[r(B^2{}_2 - B^0{}_2)\right] &= \ell\left(\Omega^2{}_2 - \Omega^0{}_2\right) + r^2\Omega^1{}_1
\end{aligned}$$

$$+ \left(1 - \frac{p\ell}{2}\right)\left(B^2{}_2 - B^0{}_2 + \frac{r^2}{\ell}B^1{}_1\right) + \mathcal{O}_3,$$

$$\partial_1\left(rB^2{}_2\right) = \frac{p\ell}{2}B^0{}_2 + B^2{}_2 + \frac{r^2}{\ell}B^1{}_1 - \ell\Omega^0{}_2 + \mathcal{O}_3, \tag{C.1}$$

where:

$$e_\mu = \left(a + \frac{\alpha_3 p}{2}\right)\omega^0{}_\mu + \left(\alpha_4 + \frac{ap}{2}\right)b^0{}_\mu + \frac{a}{\ell}b^2{}_\mu + \frac{\alpha_3}{\ell}\omega^2{}_\mu,$$

$$m_\mu = \left(a + \frac{\alpha_3 p}{2}\right)\omega^2{}_\mu + \left(\alpha_4 + \frac{ap}{2}\right)b^2{}_\mu + \frac{a}{\ell}b^0{}_\mu + \frac{\alpha_3}{\ell}\omega^0{}_\mu.$$

From the expressions (C.1), one easily concludes that the terms $\mathcal{E}^\alpha$ and $\mathcal{M}^\alpha$, included in the surface integral (6.2) for Γ, satisfy the following asymptotic conditions:

$$\partial_1 \mathcal{E}^1 = \mathcal{O}_3, \qquad \mathcal{E}^2 = \mathcal{O}_3, \qquad \partial_1 \mathcal{M}^1 = \mathcal{O}_3, \qquad \mathcal{M}^2 = \mathcal{O}_3. \tag{C.2}$$

In a similar manner, equations (5.5b) lead to:

$$\partial_1\left(rB^0{}_0\right) = B^0{}_0 + \frac{p\ell}{2}B^2{}_0 + \frac{r^2}{\ell^2}B^1{}_1 - \ell\Omega^2{}_0 + \mathcal{O}_3,$$

$$\partial_1(re_0) = \mathcal{O}_3, \qquad \partial_1(rm_0) = \mathcal{O}_3,$$

$$\partial_2 e_0 - \partial_0 e_2 = \mathcal{O}_3, \qquad \partial_2 m_0 - \partial_0 m_2 = \mathcal{O}_3,$$

$$\ell e_0 + m_2 = \mathcal{O}_3, \qquad \ell m_0 + e_2 = \mathcal{O}_3. \tag{C.3}$$

References

[1] S. Deser, R. Jackiw and G 't Hooft, *Three-dimensional Einstein gravity: dynamics of flat space*, Ann. Phys. (N.Y) **152** (1984) 220; S. Deser and R. Jackiw, *Three-dimensional cosmological gravity: dynamics of constant curvature*, Ann. Phys. (N.Y) **153** (1984) 405; E. Martinec, *Soluble systems in Quantum Gravity*, Phys. Rev. D **30** (1984) 1198; T. Banks, W. Fischler and L. Susskind, *Quantum cosmology in 2+1 and 3+1 dimensions*, Nucl. Phys. B **262** (1985) 159.

[2] For a recent review and an extensive list of references, see: S. Carlip, *Conformal Field Theory, (2+1)-Dimensional Gravity, and the BTZ Black Hole*, Class. Quant. Grav. **22** (2005) R85-R124, preprint arXiv:gr-qc/0503022; *Quantum Gravity in 2+1 Dimensions* (Cambridge University Press, Cambridge, 1998); *Lectures on (2+1)-Dimensional Gravity*, J. Korean Phys. Soc. **28** (1995) 447.

[3] J. D. Brown and M. Henneaux, *Central Charges in the Canonical Realization of Asymptotic Symmetries: An Example from Three Dimensional Gravity*, Comm. Math. Phys. **104** (1986) 207; see also: M. Henneaux, *Energy-momentum, angular momentum, and supercharge in 2+1 dimensions*, Phys. Rev. D **29** (1984) 2766; J. D. Brown, *Lower Dimensional Gravity* (World Scientific, Singapore, 1988).

[4] E. Witten, *2+1 dimensional gravity as an exactly soluble system*, Nucl. Phys. B **311** (1988) 46; A. Achucarro and P. Townsend, *A Chern-Simons Action For Three-Dimensional Anti-De Sitter Supergravity Theories*, Phys. Lett. B **180** (1986) 89.

[5] M. Bañados, C. Teitelboim and J. Zanelli, *The Black Hole in Three-Dimensional Spacetime*, Phys. Rev. Lett. **16** (1993) 1849; M. Bañados, M. Henneaux, C. Teitelboim and J. Zanelli, *Geometry of 2+1 Black Hole*, Phys. Rev. D **48** (1993) 1506.

[6] O. Coussaert, M. Henneaux and P. van Driel, *The asymptotic dynamics of three-dimensional Einstein gravity with negative cosmological constant*, Class. Quant. Grav. **12** (1995) 2961.

[7] M. Bañados, *Global charges in Chern-Simons theory and 2+1 black hole*, Phys. Rev. D **52** (1996) 5861.

[8] A. Strominger, *Black hole entropy from Neark̃Horizon Microstates*, JHEP **9802** (1998) 009.

[9] J. Navaro–Salas and P. Navaro, *A Note on Einstein Gravity on AdS_3 and Boundary Conformal Field Theory*, Phys. Lett. B **439** (1998) 262.

[10] M. Bañados, *Three-dimensional quantum geometry and black holes*, Invited talk at the Second Meeting "Trends in Theoretical Physics", held in Buenos Aires, December, 1998, preprint arXiv:hep-th/9901148.

[11] M. Bañados, *Notes on black holes and three-dimensional gravity*, Proceedings of the VIII Mexican School on Particles and Fields, AIP Conf. Proc. No. 490 (AIP, Melville, 1999), p. 198.

[12] M. Bañados, T. Brotz and M. Ortiz, *Boundary dynamics and the statistical mechanics of the 2+1 dimensional black hole*, Nucl. Phys. B **545** (1999) 340.

[13] J. Zanelli, *Chern-Simons Gravity: From 2+1 to 2n+1 Dimensions*, Lectures at the XX Econtro de Fisica de Particulas e Campos, Brasil, October 1998, and at the Fifth La Hechicera School, Venezuela, November 1999, Braz. J. Phys. **30** (1999) 251.

[14] F. W. Hehl, *Four lectures on Poincaré gauge theory*, Proceedings of the 6th Course of the School of Cosmology and Gravitation on Spin, Torsion, Rotation and Supergravity, held at Erice, Italy, 1979, eds. P. G. Bergmann, V. de Sabbata (Plenum, New York, 1980) p. 5; E. W. Mielke, *Geometrodynamics of Gauge Fields* – On the geometry of Yang-Mills and gravitational gauge theories (Academie-Verlag, Berlin, 1987).

[15] M. Blagojević, *Gravitation and gauge symmetries* (Institute of Physics Publishing, Bristol, 2002); see also: *Three lectures on Poincaré gauge theory*, Lectures at the 2nd Summer School in Modern Mathematical Physics, Kopaonik, Yugoslavia, 1-12 September 2002, SFIN A **1** (2003) 147, preprint arXiv:gr-qc/0302040.

[16] E. W. Mielke, P. Baekler, *Topological gauge model of gravity with torsion*, Phys. Lett. A **156** (1991) 399.

[17] P. Baekler, E. W. Mielke, F. W. Hehl, *Dynamical symmetries in topological 3D gravity with torsion*, Nuovo Cim. B **107** (1992) 91.

[18] T. Kawai, *Teleparallel theory of (2+1)-dimensional gravity*, Phys. Rev. D **48** (1993) 5668; *Poincaré gauge theory of (2+1)-dimensional gravity*, Phys. Rev. D **49** (1994) 2862; *Exotic black hole solution in teleparallel theory of (2+1)-dimensional gravity*, Prog. Theor. Phys. **94** (1995) 1169; A. A. Sousa, J. W. Maluf, *Canonical Formulation of Gravitational Teleparallelism in 2+1 Dimensions in Schwinger's Time Gauge*, Prog. Theor. Phys. **104** (2000) 531.

[19] A. García, F. W. Hehl, C. Heinecke and A. Macías, *Exact vacuum solutions of a (1+2)-dimensional Poincaré gauge theory: BTZ solution with torsion*, Phys. Rev. D **67** (2003) 124016.

[20] M. Blagojević and M. Vasilić, *Asymptotic symmetries in 3D gravity with torsion*, Phys. Rev. D **67** (2003) 084032;

[21] M. Blagojević and M. Vasilić, *Asymptotic dynamics in 3D gravity with torsion*, Phys. Rev. D **68** (2003) 124007.

[22] M. Blagojević and M. Vasilić, *3D gravity with torsion as a Chern-Simons gauge theory*, Phys. Rev. D **68** (2003) 104023.

[23] M. Blagojević and M. Vasilić, *Anti-de Sitter 3-dimensional gravity with torsion*, Invited talk at Workshop on Quantum Gravity and Noncommutative Geometry, Universidade Lusófona, Lisbon, 20–23 July 2004; Mod. Phys. Lett. A **17-18** (2005) 1285, preprint arXiv:gr-qc/0412072.

[24] Yuri N. Obukhov, *New solutions in 3D gravity*, Phys. Rev. D **68** (2003) 124015.

[25] S. W. Hawking and G. F. R. Ellis, *The Large Scale Structure of Spacetime* (Cambridge University Press, Cambridge, 1973).

[26] M. Henneaux and C. Teitelboim, *Asymptotically Anti-de Sitter Spaces*, Commun. Math. Phys. **98** (1985) 391.

[27] L. Castellani, *Symmetries of constrained Hamiltonian systems*, Ann. Phys. (N.Y) **143** (1982) 357.

[28] T. Regge and C. Teitelboim, *Role of surface integrals in the Hamiltonian formulation of general relativity*, Ann. Phys. (N.Y) **88** (1974) 286.

[29] J. D. Brown and M. Henneaux, *On the Poisson bracket of differentiable generators in classical field theory*, J. Math. Phys. **27** (1986) 489.

In: Trends in General Relativity and Quantum Cosmology
Editor: Charles V. Benton, pp. 125-149
ISBN 1-59454-794-7
© 2006 Nova Science Publishers, Inc.

Chapter 7

COSMOLOGICAL PRESSURE FLUCTUATIONS AND SPATIAL EXPANSION

Dale R. Koehler
82 Kiva Place
Sandia Park, New Mexico 87047

ABSTRACT

Most recently, experimental determinations of the spectrometric characteristics and internal structural velocities of galaxies have suggested the presence of massive central black holes. The analyses of the galactic spectrometric electromagnetic frequency shifts have resulted in a correlation between the hole mass and the surrounding bulge mass. In the present work, we examine whether conditions existed in the early universe, that could have led to the formation of gravitational structures possessing such unusual characteristics. We propose an early time pressure fluctuation model, which would have generated a radiation based energy distribution possessing the characteristic of a centrally collapsed zone isolated from its surrounding environment and thereby manifesting such a black hole behavior. As a hole-core expansion model, it exhibits a time evolving matter and radiation distribution, leading to a supplementary treatment of early time cosmological energy fluctuations. Einstein's gravitational equations are assumed to apply within the radiation-dominated hole-core spatial domain and, with utilization of a spherically symmetric isotropic metric, are used in order to calculate the evolutionary time expansion characteristics. Birth times for the radiation structures are uniquely correlated with the size of the spheres and are primarily determined from the early time energy densities and the apparent curvatures presented by the gravitational equations. Pressure and temperature characteristics are calculated. The hole-core model is described as a flat metric, matter plus radiation, $\sigma = 1/3$, energy distribution. It displays an early time pressure fluctuation collapse, tentatively interpreted to be the formation of a galaxy hole, and therein provides a theoretical basis for the experimental data.

PACS: 04.70.Bw, 98.80.Hw

1. INTRODUCTION

In the early part of the century, Hubble and Humason [1] cataloged the recessional velocities, relative to our Milky Way galaxy, of numerous galaxies, thereby identifying the velocity versus distance relationship now known as the Hubble law. Recessional velocities were calculable from analyses of the spectroscopic data associated with the individual light sources while distances were determined from the apparent magnitudes of these sources. It is now believed that these optical Doppler frequency shifts arise from a cosmological expansion of the intervening space itself, Einstein [2], rather than from a motion of the galaxies through space, Silk [3], Ferris [4]. More recently the experimental measurements have been extended to examine the internal galactic structure with results that suggest the presence of a massive black hole central to most, if not all, observable galaxies, Kormendy [5], Magorrian [6], Kormendy [7] and Gebhardt [8].

Did conditions exist in the early universe which could have led to the formation of gravitational structures possessing such unusual characteristics? In the present work, we examine an early time pressure fluctuation model, which would have generated a radiation based energy distribution possessing the characteristic of a centrally collapsed zone isolated from its surrounding environment and thereby exhibiting such a black hole behavior. To further explore this description, pressure fluctuations with energy densities simulating experimentally observed hole and bulge masses have been investigated. The galaxy-core model, as developed, leads to energy densities and formation times when radiation was the dominant cosmological energy contributor. The birth, or formation, times are less than a year, even for the largest energy considered which was equivalent to approximately 10^{12} solar masses. At these early times the radiation energy dominates by a factor of several orders of magnitude over mass energy densities. Since we are considering structural birth times in the range of $(10)^{-4}$ to $(10)^{-1}$ years, cosmological expansion factors are in the range of $(10)^{-8}$ to $(10)^{-6}$ with temperatures in the range of $(10)^{8}$ K and radiation densities of the order of $(10)^{2}$ to $(10)^{-3}$ grams/cm^3.

2. CORE AND HOLE MODELING

The galaxy-core model begins with an Einsteinian gravitational treatment of the early matter and radiation density fluctuation defining the core. Einstein's gravitational equations, Einstein [9], when qualified on a cosmological scale to an isotropic homogeneous space, have been solved by Friedmann, Einstein [9], to yield the classic expanding universe equations for a perfect fluid, or continuous matter distribution;

$$G_{ab}(g_{ab}) \equiv R_{ab} - \frac{1}{2}g_{ab}R = 8\pi T_{ab} = 8\pi[\rho u_a u_b + p(g_{ab} + u_a u_b)] \quad . \quad (1)$$

G_{ab} is the Einstein tensor, a function of the metric g_{ab} and its first two derivatives, R_{ab} is the Ricci tensor, R the Ricci scalar, and T_{ab} the energy-momentum tensor describing the material contents of the environment; ρ = total energy density and p = *constant* x ρ = $\sigma\rho$ =

pressure energy density. The commonly used Robertson-Walker metric, for the three curved spaces is

$$ds^2 = -a(t)^2\left(1+\frac{kr^2}{4}\right)^{-2}\left[dr^2 + r^2 d\Omega^2\right] + dt^2, \text{ where}$$
$$k = +1, 0, -1, \quad (2)$$

for a sphere, plane or pseudosphere. The Friedmann solutions to these equations are

$$\frac{\ddot{a}}{a} = -\frac{4\pi}{3}(\rho + 3p) \quad \text{and} \quad \left(\frac{\dot{a}}{a}\right)^2 = \frac{8\pi}{3}\rho - \frac{k}{a^2}, \quad (3)$$

where a is the time evolving spatial expansion parameter and k is the space curvature parameter.

As a similar model for the present calculations, the galaxy-core has been described in evolutionary terms as a continuous matter and radiation density fluctuation with $\sigma = 1/3$, with a flat space metric (producing an associated radius dependent attraction or curvature), and beginning at time $t_{core\text{-}birth}$. This is an evolving spatial and matter region, which subsequently begins collapsing in the central core zone, a temporal diminution of the hole. However, describing this spherically symmetric, radial coordinate centered, pressure fluctuation region requires a modified metric, which in the present treatment employs the isotropic form utilized by Tolman [10],

$$ds^2 = g_{11}\left[dr^2 + r^2 d\Omega^2\right] + g_{44} dt^2 = $$
$$= -e^{\mu}\left[dr^2 + r^2 d\Omega^2\right] + e^{\nu} dt^2, \quad (4)$$

where

$$\mu = \mu(r,t) \quad \text{and} \quad \nu = \nu(r,t).$$

The system of equations represented by equation (1) and as calculated by Tolman [10], is shown more explicitly in equation (5);

$$8\pi T_1^1 = -e^{-\mu}\left[\frac{\mu'^2}{4} + \frac{\mu'\nu'}{2} + \frac{\mu'+\nu'}{r}\right] + e^{-\nu}\left[\ddot{\mu} + \frac{3}{4}\dot{\mu}^2 - \frac{\dot{\mu}\dot{\nu}}{2}\right],$$

$$8\pi T_2^2 = -e^{-\mu}\left[\frac{\mu''}{2} + \frac{\nu''}{2} + \frac{\nu'^2}{4} + \frac{\mu'+\nu'}{2r}\right] + e^{-\nu}\left[\ddot{\mu} + \frac{3}{4}\dot{\mu}^2 - \frac{\dot{\mu}\dot{\nu}}{2}\right] = 8\pi T_3^3,$$

$$8\pi T_4^4 = -e^{-\mu}\left[\mu'' + \frac{\mu'^2}{4} + \frac{2\mu'}{r}\right] + e^{-\nu}\left[\frac{3}{4}\dot{\mu}^2\right],$$

$$8\pi T_4^1 = +e^{-\mu}\left[\dot{\mu}' - \frac{\dot{\mu}\nu'}{2}\right],$$

$$8\pi T_1^4 = -e^{-\nu}\left[\dot{\mu}' - \frac{\dot{\mu}\nu'}{2}\right]. \quad (5)$$

We restrict the present modeling to the case of no radial energy flow thereby requiring that

$$\dot{\mu}' = \frac{\dot{\mu} v'}{2}. \tag{6}$$

The static solutions and a general dynamic solution to this equation are

1. $\dot{\mu} = 0$,
2. $\mu = f(r) + g(t)$; $v = v(t)$ or $const$, and
3. $\mu(r,t) = \mu_0 e^{v(r,t)/2} + \mu_1$; $v(r,t) = \varepsilon(r) + \lambda(t)$. $\tag{7}$

The usual notation, where primes denote differentiation with respect to the radial coordinate r and dots denote differentiation with respect to the time coordinate t, is employed. Static solution 1, Schwarzschild's interior and exterior solutions for the case of an incompressible perfect fluid sphere of constant density surrounded by empty space, is provided in Tolman [10]. A zero-pressure surface-condition and matching and normalization of the interior and exterior metrics at the sphere radius were used as boundary conditions. Here we use the general solution form 3 of equation (7),

$$\mu(r,t) = \mu_0 e^{v(r,t)/2} + \mu_1,$$
$$v(r,t) = \varepsilon(r) + \lambda(t). \tag{8}$$

μ_1 is either a constant or a function of r, that is $\mu_1(r)$. The radiation, or perfect fluid, character of the model requires that

$$T_1^1 = T_2^2 \quad \text{and}$$
$$T_1^1 = -\frac{1}{3} T_4^4. \tag{9}$$

From equation set (5) then,

$$-e^{-\mu}\left[\frac{\mu'^2}{4} + \frac{\mu' v'}{2} + \frac{\mu' + v'}{r}\right] = -e^{-\mu}\left[\frac{\mu''}{2} + \frac{v''}{2} + \frac{v'^2}{4} + \frac{\mu' + v'}{2r}\right],$$

which must hold true for either static or dynamic conditions, and after substituting solution form (8) we get

$$-v''\left[\frac{\mu_0}{2} e^{v/2} + 1\right] - \frac{v'^2}{2}\left[\frac{\mu_0}{2} e^{v/2} + 1 - \frac{\mu_0}{2} e^{v/2}\left(\frac{\mu_0}{2} e^{v/2} + 2\right)\right] +$$
$$+ v'\left[\left(\mu_1' + \frac{1}{r}\right)\left(\frac{\mu_0}{2} e^{v/2} + 1\right)\right] = -\mu_1'' + \frac{\mu_1'^2}{2} + \frac{\mu_1'}{r}. \tag{10}$$

We choose the constant form for μ_1 which leads to

$$v'' + \frac{v'^2}{2}\left[1 - \frac{\mu_0}{2}e^{v/2}\frac{\left(\frac{\mu_0}{2}e^{v/2} + 2\right)}{\left(\frac{\mu_0}{2}e^{v/2} + 1\right)}\right] - \frac{v'}{r} = 0. \tag{11}$$

Under quasi-static conditions at time t = 0, we make the substitution $x(r, t=0) \equiv x(r, 0) = \mu_0 e^{v/2}/2$ and get

$$x' = A_0 r(1+x)e^{1+x} \text{ and}$$

$$\ln(1+x) + \sum_{k=1}^{\infty}\frac{[-(1+x)]^k}{k*k!} = A_0 \frac{r^2}{2} + A_1. \tag{12}$$

The metric associated x function, the left side of equation (12), approaches a constant (−0.797) as x approaches zero and interestingly approaches the limit value −C, where C is the "Euler-Mascheroni constant" (0.577), as x approaches ∞. C is defined as

$$C \equiv \lim_{n \to \infty}\left(\sum_{k=1}^{n}\frac{1}{k} - \ln(n)\right).$$

Examining the large r behavior, we demand that the metric $g_{44} = e^v$ approach the flat space limit $g_{44} = 1$. For large r, x approaches minus 1 and the x function approaches $ln(1+x)$, therefore

$$1 + x \approx e^{A_0\frac{r^2}{2} + A_1}, \text{ or with } A_0 \text{ negative,}$$

$$\mu_0 e^{v/2} \approx -2 \text{ and } \mu_0 = -2; \quad x = -e^{v/2}. \tag{13}$$

Similarly, $g_{11} = -e^\mu$ is also set equal to −1 at large r. Since μ is $\mu_1 + \mu_0 * e^{v/2} = \mu_1 - 2 * e^{v/2}$, then $\mu_1 = 2$ and $g_{11} = -e^{2(1+x)}$.

At this juncture only the requirement of no radial energy flow and large r asymptotic agreement with a flat space form has been demanded. The large r behavior requirement is obviously applicable to the exterior solution, however the character of the space and radiation interior to the sphere in question is not dissimilar to the exterior region. The interior and exterior metrics are therefore considered formally equivalent with only normalization forthcoming.

We now introduce the singular physical requirement, also posed by Tolman in his static description of a constant density sphere embedded in empty space, to uniquely describe the pressure fluctuation, namely that the pressure at the sphere's radius go to zero. We have then a spherical bubble of radiation embedded in a surrounding sea of radiation with the only characterizing distinction being a radial zero in the pressure distribution at the sphere radius.

Reference is now made to equation set (5) where the pressure is given by

$$8\pi T_1^1 = -e^{-\mu}\left[\frac{\mu'^2}{4} + \frac{\mu'\nu'}{2} + \frac{\mu'+\nu'}{r}\right] + e^{-\nu}\left[\ddot{\mu} + \frac{3}{4}\dot{\mu}^2 - \frac{\dot{\mu}\dot{\nu}}{2}\right] =$$
$$= -8\pi \; pressure. \tag{14}$$

We restrict ourselves to the static portion of the equation since in any event we will use the static description as the starting point for the time development. In the x notation we have

$$8\pi \; pressure = e^{-2(1+x)}\frac{x'}{x}\left[x'(x+2) + \frac{2}{r}(x+1)\right]. \tag{15}$$

For the pressure = 0 condition, we get

$$0 = e^{-2(1+x)}\frac{A_0 r(1+x)e^{1+x}}{x}\left[A_0 r(1+x)e^{1+x}(x+2) + \frac{2}{r}(x+1)\right]. \tag{16}$$

This equation sets the constant A_0 to

$$A_0(x_1, r_1) = -\frac{2}{r_1^2}\frac{e^{-(x_1+1)}}{x_1+2} \tag{17}$$

where $x_1 = x(r_1, 0)$ and r_1 is the sphere radius. The solution afforded by setting $x_1 = x(r_1, 0) = -1$ is rejected since it leads to requiring $A_1 = \infty$. As mentioned in the discussion of the asymptotic behavior of the metric in equation (13), A_0 satisfies the required negative character for the integration constant.

Equation (5) for the energy density T_4^4, although at present utilized for a flat space environment, exhibits a curvature energy component. The radiation energy density and this important curvature energy component determine the time evolution of the radiation space, both interior and exterior to the radiation sphere. We rewrite equation (5) in the latter-day expansion factor notation where e^{μ} is set equal to a^2, a being the expansion factor, to illustrate a comparison to the cosmological form:

$$8\pi T_4^4 = -\frac{1}{a^2}\left[2\frac{a''}{a} - \left(\frac{a'}{a}\right)^2 + \frac{4}{r}\frac{a'}{a}\right] + \frac{3}{1-\ln(a^2)}\left(\frac{\dot{a}}{a}\right)^2 \quad or$$

$$8\pi T_4^4 \equiv "curvature" + \frac{3}{1-\ln(a^2)}\left(\frac{\dot{a}}{a}\right)^2. \tag{18}$$

This expansion rate equation is to be compared with Friedmann's cosmological expansion rate equation (3). From the definition of μ and x, that is, $\mu = 2(1+x)$, we also see that since e^{μ} is associated with a^2, then $d\mu/dt = 2*dx/dt = 2*a^{-1}*da/dt$ and $dx/dt = a^{-1}*da/dt$, the Hubble factor.

In the x notation, the curvature is given by

$$curvature = -e^{-2(1+x)} x' \left[x'\left(\frac{5+3x}{1+x}\right) + \frac{6}{r} \right]. \quad (19)$$

From the energy density equation of the equation set (5), we can write with the x notation,

$$\left(\frac{\dot{x}}{x}\right)^2 = \frac{8\pi\rho(t)}{3} - \left[-\frac{e^{-2(1+x)}}{3}\left(x'^2\frac{5+3x}{1+x} + 6\frac{x'}{r}\right)\right] =$$

$$= \frac{8\pi\rho(t_{core-birth})}{3} e^{-4(1+x)} - \left[-\frac{e^{-2(1+x)}}{3}\left(x'^2\frac{5+3x}{1+x} + 6\frac{x'}{r}\right)\right]. \quad (20)$$

Equation (20) represents an energy balance statement in that the difference between the radiation and/or matter energy and the warping, or spatial curvature energy, goes into the energy of spatial expansion. Radiation energy density $\rho(t)$ has been represented more explicitly in terms of the time evolution parameter given by $e^{-4(1+x)}$ and subsequently, as an initial condition, will be expressed in terms of the present day radiation energy density, Ω_{r0}. Initial conditions are such that in those regions where the curvature energy density is positive and exceeds the radiation energy density, then $(dx/dt)^2$ is negative and a collapsing space condition will ensue. With an initial constant radiation energy density and, as a result of the radial behavior of the curvature component, the region internal to the collapse boundary is cut off from the expanding external region since the temporal expansion rate, dx/dt, goes to zero at this boundary (zero propagation velocity). In addition, there exists throughout the immediate proximal region of the sphere a variable propagation velocity since the null geodesic produces a propagation velocity equal to

$$propagation\ velocity = \sqrt{\frac{e^\nu}{e^\mu}}. \quad (21)$$

In the static solution of Schwarzschild and the isotropic static solution of Tolman, where $e^\nu = ((1 - r_s/r)/(1 + r_s/r))^2$ and $e^\mu = (1 + r_s/r)^4$, the propagation velocity goes to zero at the Schwarzschild radius r_s. As a boundary condition in the present model, the radius of the collapse zone, referred to as the hole region, and the radial zero of the propagation velocity (at $x(r_{hole}) = -e^{\nu/2} = 0$), are set equal thus providing the normalization referred to earlier. This boundary condition determines A_1 as

$$A_1 = \ln(1) + \sum_{k=1}^{\infty} \frac{-1^k}{k*k!} - A_0 \frac{r_{hole}^2}{2}. \quad (22)$$

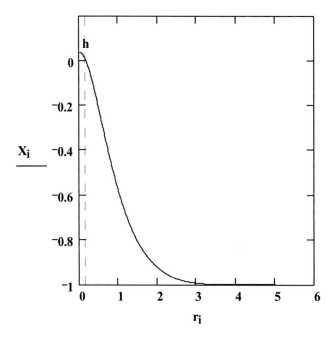

Fig. 1. Metric x-factor diagram as a function of the spherical radial coordinate (in units of the radiation sphere radius, r_1) illustrating the zero at the hole radius h.

The interior and exterior metrics are equal at the sphere's radial boundary but not equal to the static Schwarzschild solution values, where the exterior solution is for empty space; the empty-space static-solution values are again,

$$g_{44}(r_1) = \left(\frac{1 - r_s/r_1}{1 + r_s/r_1}\right)^2 \text{ and } g_{11}(r_1) = -(1 + r_s/r_1)^4. \tag{23}$$

With the values of A_0 and A_1 thus determined, $x(r,0)$, moreover, is now provided and, in particular, the value of $x(r_1,0)$ is fixed (see equation (12)) when evaluated at the boundary r_1,

$$\ln(1 + x_1) + \sum_{k=1}^{\infty} \frac{[-(1 + x_1)]^k}{k * k!} = A_0(x_1, r_1)\frac{r_1^2}{2} + A_1(x_1, r_1). \tag{24}$$

Figure 1 displays the x function versus the radial coordinate r with the transition from positive to negative at the hole radius. The metric quantity, x_1, will be necessary for a subsequent calculation of birth times. The solution for $x_1 = x(r_1,0)$ is

$$\ln(1 + x_1) + \sum_{k=1}^{\infty} \frac{[-(1 + x_1)]^k}{k * k!} = -\frac{e^{-(x_1+1)}}{r_1^2(x_1 + 2)}\left(r_1^2 - r_{hole}^2\right) + \ln(1) + \sum_{k=1}^{\infty} \frac{-1^k}{k * k!}. \tag{25}$$

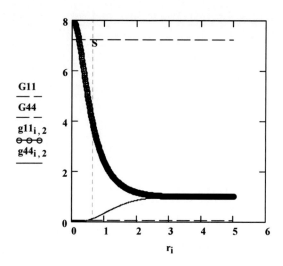

Fig. 2. Model metrics: the spatial, $g_{11,2} = -g_{11} = e^{\mu}$, and the temporal, $g_{44} = e^{\nu}$, metrics are illustrated as a function of the radial coordinate r (in units of the radiation sphere radius, r_1); the subscript 2 indicates usage of the experimental hole-mass value. The Schwarzschild radius, S, and the static-empty-space solutions, G11(negative) and G44, are shown for reference.

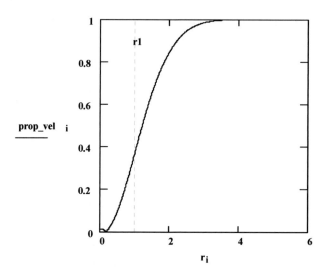

Fig. 3. Propagation velocity as a function of the radial coordinate r (in units of the radiation sphere radius, r_1).

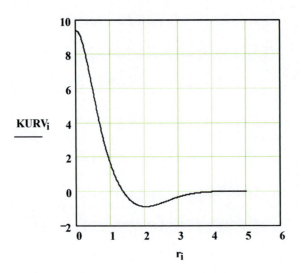

Fig. 4. Curvature, on a linear scale, as a function of the radial coordinate r (in units of the radiation sphere radius, r_1). Curvature is in units of $(10)^{-30}/cm^2$.

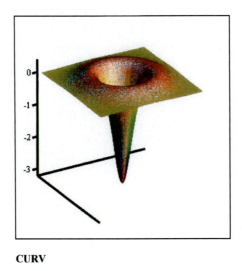

CURV

Fig. 5. Two dimensional representation of the three dimensional curvature function. Negative values of the displayed function represent positive curvatures.

Cosmological Pressure Fluctuations and Spatial Expansion 135

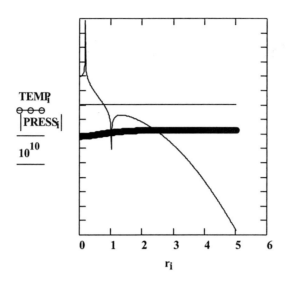

Fig. 6. Pressure and temperature (K) throughout the radiation sphere's proximity. The radial coordinate is in units of the radiation sphere radius r_1. Pressure is in units of $(10)^{-40}/\text{cm}^2$. The reference value, $(10)^{10}$, is indicated.

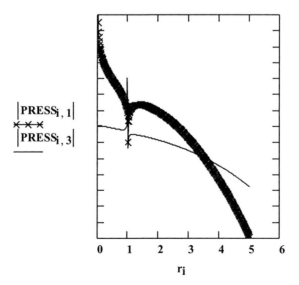

Fig. 7. Pressure function versus radius for the case of the hole mass equal to the experimental value (subscript 1) and for the hole radius equal to the sphere radius (subscript 3) where the singularity at the hole boundary has merged with the zero at the sphere radius. Radial coordinate in units of the sphere radius r_1.

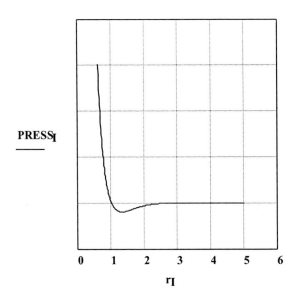

Fig. 8. Pressure in the immediate proximity of the radius of the sphere. A linear pressure scale showing the behavior at the pressure zero. The radial coordinate is in units of the radiation sphere radius r_1.

We choose to make the hole mass and thus the hole (collapsed zone) radius an experimentally adjusted parameter of the model. We will, however, subsequently theoretically posit that the formation process is reversible, or quasi-static, and therefore require that, during sphere formation, the entropy change $\Delta S = 0$. This requirement leads to a theoretical determination of the hole radius.

The metrics thereby determined are displayed in Figure 2 as a function of the radial coordinate r. We have also included, for reference purposes, the Schwarzschild metric boundary values. Figure 3 illustrates the significant features of the propagation velocity where the asymptotic approach to c, at large r, is evident along with the zero at the singularity surface. Curvature, temperature and pressure plots are shown in Figures 4, 5 and 6. We have used a hole mass to sphere mass ratio of $5.2(10)^{-3}$ in these calculations in anticipation of utilizing an experimentally determined quantity as an input to the model. Curvature exhibits a positive region, at small values of the radial coordinate, transitioning to a negative region and subsequently approaching zero (flat space) at large r. The pressure plot displays the zero introduced as a model condition at the sphere boundary and a singularity at the hole boundary caused by the zero metric normalization. Pressure inside the hole region is large and negative and at the hole edge approaches the singularity. Between the hole edge and the sphere radius, the pressure is positive and rapidly decreasing to zero at the sphere radius. External to the sphere, the negative pressure is small and decreases to zero at large radial distances. For some conditions of density and curvature, the hole region increases and when the hole radius approaches the sphere radius, the pressure singularity at the hole edge and the zero of the pressure function at the sphere radius merge at the sphere radius as illustrated in Fig. 7. Figure 8 shows, on a linear scale, the pressure factor near the zero at the sphere's radius.

Again it is instructive to compare these results with the earlier static solutions. In the Schwarzschild treatment, as constructed by Tolman, of a constant density sphere embedded in flat empty space, the curvature energy is zero in the exterior region. Tolman uses the standard form of the spherically symmetric metric and calculates the interior solution to be

$$ds^2 = -e^\lambda dr^2 - r^2 d\Omega^2 + e^\nu dt^2 =$$

$$ds^2 = -\frac{1}{1-\frac{r^2}{R^2}} dr^2 - r^2 d\Omega^2 + \left[A - B\sqrt{1-\frac{r^2}{R^2}} \right]^2 dt^2$$

$$\text{with } A = \frac{3}{2}\sqrt{1-\frac{r_1^2}{R^2}}, \quad B = \frac{1}{2} \quad \text{and} \quad R^2 = \frac{r_1^3}{r_s}. \tag{26}$$

Although not calculated there, the interior curvature component is

$$curvature = -e^{-\lambda}\left[\frac{\lambda'}{r} - \frac{1}{r^2}\right] - \frac{1}{r^2} = \frac{3}{R^2}. \tag{27}$$

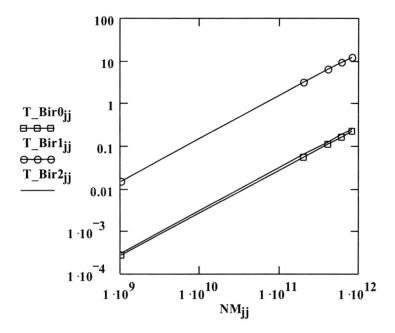

Fig. 9. Radiation sphere birth times (years) versus mass of the radiation sphere. T_Bir0 is for a zero hole-radius extremum, T_Bir1 is from the experimental hole-mass datum and T_Bir2 for a hole-radius equal to the sphere radius. Radiation sphere mass values extend from 10^9 to 10^{12} stellar masses.

If one assumes that present day galaxy holes derive from such a collapse process and that the present day hole mass (neglecting accretion processes) represents the current collapsed radiation mass, then for the two region model, calculation of the bounds on the galaxy-core birth time are forthcoming from consideration of the galaxy-core mass radius and the hole-mass singularity-radius. If the initial mass distribution's maximum radius at time $t_{core-birth}$ is less than or equal to the singularity radius for that mass, then all of the region will eventually collapse but the collapse process and the mass distribution itself will be subsequently unobservable to regions outside the singularity radius. This case corresponds to the density being equal to the curvature value at the sphere's edge. Model cases with smaller masses or earlier galaxy birth times will additionally contain the second, more rapidly expanding, negatively curved region. Collapse, as used here, describes the reduction of the time metric $x^2 = e^v$ toward zero. The spatial metric, $e^\mu = e^{2(1+x)}$, approaches e^2 as x approaches zero. Therefore, for the region internal to the singularity surface, time tends toward zero. The hole then is a region of unchanging spatial character and with no time character.

The time-bound (birth) calculation sets the collapsed mass radius, which equals $r_1 * $ *(hole mass ratio)*$^{1/3}$, equal to the sphere radius. This upper time bound is calculated to be $1.5(10)^{-2}$ years, for a galaxy-core mass of 10^9 stellar masses using equations (30) and (31) for the space expansion factor for a "radiation plus matter" universe and 0.52% for the hole-mass/core-mass ratio (present day mass density $\Omega \approx 0$ relates the galaxy-core radius to the galaxy-core mass). For mass and birth-time combinations greater than the upper extremum, the present day collapsed mass value would be greater than the experimental hole value referred to above. The calculational result for the lower time bound for the galaxy-core birth time (resulting for a galaxy-core mass radius that approaches a singularity radius equal to zero) is $2.7(10)^{-4}$ years. For cosmological times less than this bound, the radiation energy density exceeds the curvature energy density and the spheres experience no collapsing region. This dependence of birth times on galaxy-core mass is illustrated in Fig. 9.

The basic physical concept of causality requires that a coherency time, or formation time, be imposed on the time origin of these structures. Although the radiation energy density can be calculated as a function of time and associated with the time of birth of the radiation spheres, the coherency time interval must additionally be imposed to determine actual birth times. Such a coherency time interval is expressed as

$$t_{coherency} = n_{coherency} \frac{r_1}{c} . \qquad (28)$$

The coherency time interval is a measure of the time necessary to propagate formation information across the radial extent of the spherical radiation energy distribution and thereby establish equilibrium throughout. Given an instant in time when such structures would have been born, the coherency time interval determines the additional time increment to establish the starting time, or birth time, for subsequently calculating the evolutionary character of the radiation sphere. The uncertainty in expressing such a time is contained in the undetermined constant, $n_{coherency}$. At present no a priori deductive determination of this factor, however, has been forthcoming and so we proceed with the defining equations of the model and merely determine this constant as a companion calculation to be compared for reasonableness with the galaxy-core birth-time calculations. An interpretation of the fact that the experimental hole mass birth-time is so near to the lower time bound birth-time (as shown in Fig. 9) is that the radiation spheres being thus formed require only a minimum coherency time interval to begin the evolutionary contraction (hole region) and expansion (outer shell region) process. The lower time-bound birth-time can be written as

$$t_{core\ birth} = \left(\frac{3}{f0}\right)^{1.5} \frac{GM}{c^3}, \text{ where } f0 = -\left[x'(0)^2 \frac{5+3x(0)}{1+x(0)} + 6\frac{x'(0)}{r(0)}\right]\frac{1}{g_{11}(0)} \approx 2.1,$$

$$\text{or } t_{core\ birth} \approx 0.85\frac{r_s}{c}\ ;\ x'(0) = -2\sqrt{\frac{g_{44}(r(0))}{g_{44}(r_1)} \frac{r(0)}{r_1^2} \frac{1+x(0)}{2+x(r_1)}}. \tag{29}$$

f0 is a measure of the curvature function evaluated at the metric zero or hole radius.

The radius of the structure in question is a function of the mass and the density at the time *t* of formation, *t_core-birth*, and is given by equations (30);

$$R_{sphere} = r_1 = \left(\frac{3M_{sphere}}{4\pi\rho(t)\rho_c}\right)^{1/3}, \text{ where}$$

$$\rho(t) = \Omega r0\left(\frac{a_0}{a}\right)^4 = \Omega s0\left(\frac{3}{4}\frac{age}{t}\right)^2. \tag{30}$$

We have made a distinction between the time of creation and the time of core-birth, where the time of core-birth is equal to the time interval of coherency or formation after the time of creation. The density is changing during this time interval and we use the density at the core-birth time for the model calculations. Expressing Friedmann's equation (5) for the time development of the cosmological expansion factor *a*, for a two species (dust and radiation) flat universe, we write

$$\left(\frac{\dot{a}}{a}\right)^2 = \frac{8\pi\rho_{cos\ mo}}{3} - \frac{k}{a^2} = \frac{8\pi\rho_c}{3}G\left[\Omega s0\left(\frac{a_0}{a}\right)^3 + \Omega r0\left(\frac{a_0}{a}\right)^4\right] \text{ where}$$

$$\rho_c = \frac{3}{8\pi G}H_0^2 \text{ and } age = \frac{2}{3}\frac{1}{H_0} \text{ or } \rho_c = \left(6\pi G\ age^2\right)^{-1};$$

$$Temperature = T_{exp}\left(\frac{a_0}{a}\right). \tag{31}$$

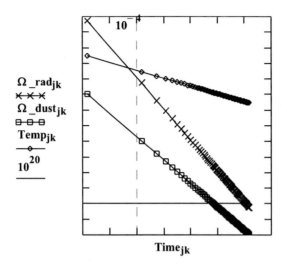

Fig. 10. Cosmological radiation and matter (dust) densities and temperature as a function of time (years) in the early universe. The dominance of the radiation energy density over the matter energy density at these early times is noted. Cosmological temperatures, multiplied by a factor of $(10)^{20}$ for display purposes, are near 10^8 K while the radiation energy density is $1.4(10)^{27} \rho_c$ at $2.7(10)^{-4}$ years.

The present day values for the radiation energy and the matter energy (approx. total energy) are written as $\Omega r0$ and $\Omega s0$ respectively. In integral form we have

$$\int_a^{a_0} \frac{a\,da}{\sqrt{A+Ba}} = \int_t^{age} C\,dt \text{ , where } A = \Omega r0(a_0)^4,\ B = \Omega s0(a_0)^3 \text{ and } C = \sqrt{\frac{8\pi\rho_c G}{3}}. \quad (32)$$

The solution is

$$t = age\left[\frac{-\left(2\dfrac{\Omega r0}{\Omega s0} - \dfrac{a}{a_0}\right)\sqrt{\dfrac{\Omega r0}{\Omega s0} + \dfrac{a}{a_0}} + 2\left(\dfrac{\Omega r0}{\Omega s0}\right)^{1.5}}{-\left(2\dfrac{\Omega r0}{\Omega s0} - 1\right)\sqrt{\dfrac{\Omega r0}{\Omega s0} + 1} + 2\left(\dfrac{\Omega r0}{\Omega s0}\right)^{1.5}}\right]. \quad (33)$$

M_{sphere} is the mass equivalent of the total radiation energy in the sphere and ρ_{cosmo} is the total cosmological energy density. $\rho_{cosmo} = \rho(t_{core\text{-}birth})$ provides the initial density condition for the radiation sphere. We have used the classical notation of equations (2), (3) and (18) for the cosmological space expansion factor a. Before time $t_{core\text{-}birth}$ the universe is assumed governed by the expansion, space density and temperature behavior determined by this two species characterization, evolving temporally according to the two-component expansion-factor equation (31). At small a, or early times, this is approximately,

$$t = age\left(\frac{a}{a_0}\right)^2 \frac{3}{4}\left(\frac{\Omega s0}{\Omega r0}\right)^{1/2}, \quad (34)$$

thereby producing equation (30). At times near the birth times of the structures considered, we have plotted in Fig. 10 the expansion factor a, the cosmological density ρ and the radiation temperature. These cosmological values determine the initial conditions for the radiation-sphere time-development equation (20).

The coherency time is now set equal to the time t, or $t_{core\text{-}birth}$, of equation (30). Since at the core edge, the time metric factor, x, goes to zero, we can use equations (12), (20), (25), (30) and (31) to solve explicitly for $n_{coherency}$ in terms of the remaining undetermined quantity, r_{hole}, or m_{hole};

$$m_{hole}/M_{sphere} = 5.2(10)^{-3}\ ;\quad x_1 = x_1(r_{hole}/r_1) = -0.579\ ;$$

$$n_c \equiv n_{coherency} = \frac{1}{4}(x_1+2)e^{x_1+1}\left[-\frac{5}{3}\left(\frac{r_{hole}}{r_1}\right)^2 + (x_1+2)e^{x_1}\right]^{-.5} = 0.626\ . \quad (35)$$

The hole mass to sphere mass ratio is an experimentally measured quantity; Kormendy [5], Magorrian [6], and Kormendy [7] (also see Fig. 2 in Gebhardt [8]). From Magorrian, log (M_{hole}/M_{bulge}) = -2.28 (mean with std.dev. = 0.51) or $M_{hole}/M_{bulge} = 5.2\times10^{-3}$. We equate our sphere mass to M_{bulge}. The coherency constant varies from approximately 0.6 to 2.2 for collapse zone radii from zero to r_1. Objectively, these values seem reasonable, thereby supporting the core-birth time values as physically and calculationally defensible.

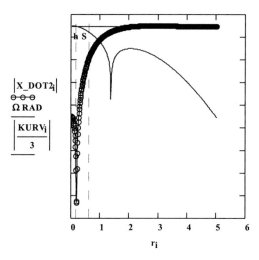

Fig. 11. Metric rate change (X_DOT2), curvature ($KURV$) and energy density factor (ΩRAD), all on a logarithmic scale, are displayed as a function of the radial coordinate in units of the sphere radius r_1. The hole radius, h, and the Schwarzschild radius, S, are indicated. ΩRAD is $2.9(10)^{-30}/\text{cm}^2$. The crossover from positive curvature dominance to radiation energy dominance creates the hole region near the radial origin.

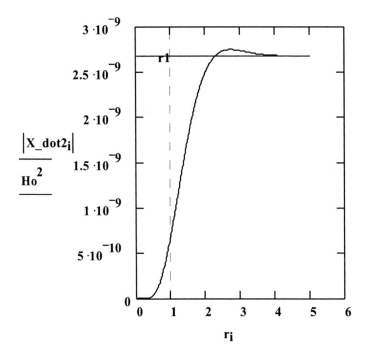

Fig. 12. Metric rate change, $(dx/dt)^2 = X\_dot2$, on a linear scale, showing comparison to the background Hubble factor $H_0^2(\sec^{-2})$.

In Fig. 11 we display the sphere density and curvature and the resultant velocity-factor behavior, as a function of r, while incorporating the experimental hole mass value in the calculations. Also shown in Fig. 12 is the character of the velocity-factor solution, at the birth time of the sphere, relative to the background radiation Hubble factor, Ho^2. The velocity-factor behavior, mentioned earlier as a measure of the energy balance, shows a resultant overall decrease for the spatial expansion process in the physical vicinity of the sphere. No energy from the embedding space has been provided to the spheres during sphere and hole formation if the entropy condition is satisfied but spatial expansion has slowed.

Rewriting equation (5) in the x notation, after incorporating the equation of state relationship between radiation pressure and radiation density, we have

$$8\pi T_1^1 = -8\pi \, pressure = e^{-\mu}\left[\frac{\mu'^2}{4} + \frac{\mu'\nu'}{2} + \frac{\mu'+\nu'}{r}\right] - e^{-\nu}\left[\ddot{\mu} + \frac{3}{4}\dot{\mu}^2 - \frac{\dot{\mu}\dot{\nu}}{2}\right] =$$

$$= -\frac{8\pi T_4^4}{3} = \frac{e^{-\mu}}{3}\left[\mu'' + \frac{\mu'^2}{4} + \frac{2\mu'}{r}\right] - e^{-\nu}\left[\frac{1}{4}\dot{\mu}^2\right]. \quad (36)$$

Or, with the aid of the definitions in equations (15) and (19), we get

$$\left(\frac{\ddot{x}}{x}\right) = \frac{8\pi\rho}{3}(1-2x) - pressure\frac{x}{2} - curvature\left(\frac{1}{3} - \frac{x}{2}\right). \quad (37)$$

Cosmological Pressure Fluctuations and Spatial Expansion 143

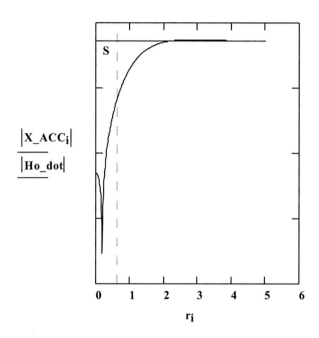

Fig. 13. Metric-rate rate-change, $d^2x/dt^2 + (dx/dt)^2 = X\_ACC$ on a logarithmic scale, showing comparison to the background Hubble-factor rate-change $dH_0/dt = Ho\_dot$. The Schwarzschild radius, S, is indicated.

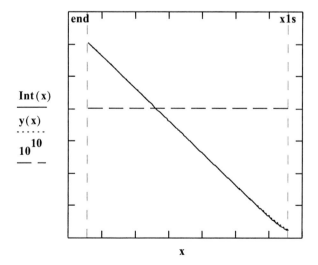

Fig. 14. Metric rate change integral. The x function integrand, from equation (38), is shown over an x range from "x0" = $x1s = x_{start}$ to "end" = $(10)^{-6} x_{start}$. In this example the integrand is accurately approximated by the $y(x)$ function $y(x) = a/x + b$, with $a = 2.1(10)^4$ and $b = -4(10)^4$.

Similarly to Figures 11 and 12, we calculate and display in Fig. 13 the results for the acceleration factor, $d^2x/dt^2 + (dx/dt)^2$. Time rate of change of the cosmological background radiation Hubble factor, dHo/dt, is compared with the acceleration factor in this figure.

Collapse begins immediately at radii less than the singularity radius. The process is a logarithmic one, however, representing the time integral of equation (20); e.g. a $(10)^6$ fold reduction in x at the central radius requires approximately $7(10)^5$ seconds when calculated with a radiation mass of $(10)^9$ stellar masses and a hole mass ratio of $5.2(10)^{-3}$. The time evolution equation, in integral form, for the metric factor x, is given by equation (38);

$$\int_0^t dt = \int_{x0}^{end} \frac{dx}{xc}\left[\frac{8\pi\rho(t_{core-birth})}{3}e^{-4(1+x)} + \frac{e^{-2(1+x)}}{3}\left(x'^2\frac{5+3x}{1+x} + 6\frac{x'}{r}\right)\right]^{-1/2}. \quad (38)$$

The x integrand and an approximating function $y(x) = a/x + b$ are plotted in Fig. 14. The approximation is excellent. Integration limits are indicated by $x0$ and $end = (10)^{-6}x0$. In this particular example, we use the radial coordinate at $r = (10)^{-5}r_1$, the experimental hole mass and a sphere mass of $(10)^9$ stellar masses. The energy density function, that is, the sum of the radiation energy density and the curvature energy density, is slowly varying as a function of x and therefore the x^2 term dominates the dependence on x, thus generating the approximate logarithmic behavior.

Reiterating, the cosmological density, $\Omega_s 0$ and the hole mass are experimental parameters. For all of the calculations for Figures 4, 6-8 and 11-14, a galaxy-core mass = 10^9 stellar masses, $\Omega_s 0 = 1$ and a Hubble's constant $H_0 = 0.485\times 10^{-10}$/year (producing an age of 13.7×10^9 years) were used. However, if, from an additional theoretical perspective, the radiation sphere formation process behaves preferentially to minimize the entropy change, then an appropriate governing entropy restriction for zero entropy change can be stated as

$$\Delta S = 0 = \int\frac{dQ}{T} = 2\int\frac{\left(4\pi r^2 e^{3\mu/2}e^{v/2}dr\right)}{T_0 e^{-\mu/2}}\left(-e^{-2(1+x)}x'\left[x'\left(\frac{5+3x}{1+x}\right)+\frac{6}{r}\right]\right). \quad (39)$$

We have used Tolman's expression (eqn. 97.2) for the total energy of the radiation sphere utilizing the perfect fluid or radiation character to express T_1^1, T_2^2 and T_3^3 in terms of T_4^4 and expressing T_4^4 as the curvature energy. The energy of formation, that is the curvature energy, is integrated over the entire spatial domain and the temperature is a function of the spatial metric $e^{-\mu/2}$; $T = T_0 e^{-\mu/2}$. A numerical integration of equation (39) yields a theoretically determined value for x_1 and therefore r_{hole}. The result is $r_{hole} = 0.72\ r_1$. At this point, it should be noted again that the gravitating radiation sphere is a rather diffuse structure where there is no density difference across the pressure = 0 boundary at r_1 nor does the curvature or distortion function exhibit any discontinuity across this same boundary. The time evolution of the sphere, deriving from spatial expansion, preserves the main spatial qualities of the sphere and it is apparent that the comparison with the experimental hole-mass to bulge-mass ratio is therefore somewhat arbitrarily defined. The ratio is calculated to vary from $(0.72)^3$ at $r_{sphere} = r_1$ to $(.72/6)^3 = 1.7(10)^{-3}$ at $r_{sphere} = 6r_1$ (see Fig.4 for the radial extent of the curvature function). Matter accretion processes occurring after sphere formation would more probably be responsible for bulge mass formation. The time-evolved radiation sphere itself then should be interpreted as the "hole structure".

What is the thermodynamic distribution function describing the expected number of radiation spheres of a given mass, or given energy content? A partitioning of the total available energy with domination by the self, or rest-mass, energy of the spheres was considered appropriate. A gas based analogy and an associated Maxwell Boltzmann distribution where

$$\delta n = nAM^3 e^{-E/kT} dE \text{ with } E = Mc^2 \text{ and}$$

$$A = c^{-2} \left(\int_0^\infty e^{-Mc^2/kT} M^3 dM \right)^{-1} = c^{-2} \left(\frac{c^2}{kT} \right)^4 (\Gamma(4))^{-1} \text{ leading to}$$

$$\delta n = \frac{n}{6} \left(\frac{E}{kT} \right)^3 e^{-E/kT} \frac{dE}{kT} \tag{40}$$

does not seem an adequate description. The energy peak in this distribution occurs at $E_{max} = 3kT$. At this time, however, the desired distribution function has not been formulated.

In chronological review, the presence at time $t_{creation}$ of an incompletely formed pressure fluctuation begins the departure from relative uniformity, which describes the early universe. A time $t_{core\ birth}$ is required for coherency or formation of the radiation spheres and is followed by both a collapsing and expanding time evolution of the distorted spatial region. In the present radiation, or perfect fluid, modeling, the description provided by equations (5) incorporates only the gravitational physics of the collapsing galaxy-core space; any subsequent matter accretion processes, or other energy sources, are not included. The continuous matter- but radiation-dominant-energy of the early universe, that constitutes the galaxy-core material, is considered the perfect fluid used to determine the expansion factor solution form. A present day experimental hole mass value, equal to 0.52% of the galaxy bulge-mass (we have associated the galaxy bulge-mass with our core-mass), has been used for the evolving hole although we have introduced an alternative theoretical entropy concept which constrains the determination of the hole radius. This radius is assumed to be the metric singularity value. The hole radius during and after the logarithmic collapse does not change while the time character of the hole collapses to zero. Temporal evolution of the galaxy-core expansion parameter, in the outer regions of the mass distribution, follows the negative curvature behavior and, asymptotically at large r, a flat (curvature = 0) behavior. Since the spatial extent of the hole region does not change, the galaxy-core density function in the hole region remains constant during the evolution. The variable $T_{exp} = T_{experimental}$ is the cosmic microwave background temperature and is utilized as the reference. Temperature is assumed to follow an a^{-1} or $e^{-\mu/2}$ behavior throughout both spatial domains.

If we calculate the radiation energy and birth time relationship for these radiation spheres (see eqn.(29)) near the cosmological inflation time, we get approximately $(10)^3$ equivalent grams (for the hole mass) at $t_{core\ birth} = (10)^{-33}$ seconds ($(10)^{-33}$ seconds is defined as the end of inflation). In this calculation, we have used the hole mass rather than the sphere mass. The contemporary attempt to relate a collapsed gravitational entity to the fundamental particle spectrum, for example baryons, therefore is seen in the present model to lead to the requirement of an additional mass reduction factor (going backward in time) of $(10)^{-27}$. Since, for these radiation spheres, the mass (from eqn. 29) is inversely proportional to the square root of the density and in the radiation environment we are considering, the density is inversely proportional to the fourth power of the expansion factor (inversely proportional to the second power of the time), a change in the expansion factor proportional to the square root of the mass factor achieves the desired reduction. A change of $(10^{-27})^{.5} = 3.2(10)^{-14}$ (fourteen orders of magnitude) in the expansion factor therefore would produce the conditions for baryon-like radiation spheres. This puts such a proton-like particle creation-time during the period of cosmological inflation since the inflation period produces an expansion factor change of the order of thirty orders of magnitude. However if the release of latent energy from the phase transition of the inflation process maintained the energy density of the universe during the transition, then further expansion factor reduction would have to have occurred before inflation. For the time period from 10^{-35} to 10^{-43} seconds (Planck time), however, sufficient reduction could not have resulted. By contrast, in the rest-energy based Maxwell Boltzmann distribution of equation (39), distributions of radiation spheres with holes approximating baryon masses would peak ($3kT$) at a temperature of approximately $7.1(10)^{14}$K which occurs at about $6.6(10)^{-10}$ seconds. At the post inflation time of $(10)^{-33}$ seconds, the temperature is $5.8(10)^{26}$K representing a peak sphere mass of $2.7(10)^{-13}$ grams and an associated hole mass of $1.4(10)^{-15}$ grams.

3. OBSERVATIONAL REDSHIFT MODELING

For radiation emission and detection processes, photon propagation behavior along the time evolving emitter to observer path, is determined by the equation for null geodesics and integration over time along the light path. This leads to the expression relating the emitted and observed time intervals, Linder [11],

$$\int_{t_e}^{t_{e2}} dt/a = \int_{t_o}^{t_{o2}} dt/a \; ; \; t_{e2} = t_e + \Delta t_e \; ; \; t_{o2} = t_o + \Delta t_o \; . \tag{41}$$

The resulting Friedmann-Robertson-Walker (FRW) redshift expression as in Linder [11], is

$$z = \frac{v_e - v_o}{v_o} = \Delta t_o / \Delta t_e - 1 = a(t_o)/a(t_e) - 1 \tag{42}$$

and illustrates that only the initial and final expansion states, $a_j(t_i)$ (i = emitter time or observer time, j = emitter or observer), determine the overall resultant change in frequency or wavelength of the propagating radiation. In other words, in a non-monolithic universe where local warping is present and where radiation emission sources and radiation observers (detectors) are both embedded in such locally warped regions, calculation of the radiation modification (redshift) involves calculation of the local region's expansion state as manifest in the local expansion parameter $a_j(t_i)$. The potential energy, or wavelength, diagram illustrated in Fig. 15 is a pictorial representation of the evolving photon energy state as it propagates (1), through the emitting warped galaxy region, (2), out of the warped interface, (3), through the spatially expanding path between emitting and observing galaxies, (4), into the observer space-galaxy interface, and finally, (5), to the detection point within the observer galaxy. The observer galaxy and the emitter galaxy are assumed to exhibit the same time evolutionary or expansion characteristics. In such a path, the wavelength stretching (photon energy loss) step at the emitter galaxy-space interface and the energy loss process during intergalactic travel is mirrored at the second observer space-galaxy interface where the energy loss is partially recovered and the photon wavelength decreases. The expanding emitter and observer galaxy regions therefore produce the net overall energy change, or photon wavelength increase, during the time interval from emission to detection. If the $a_{emitter}$ and $a_{observer}$ evolution lines corresponded to the same $t^{2/3}$ time behavior as a_{space}, then no energy loss or recovery would be incurred at the space-galaxy interfaces. Although the present model calculations are limited to the core expansion-factor time-development, the notion of local space warping and its impact on the propagating radiation is still appropriate.

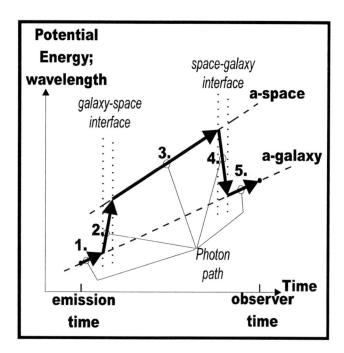

Fig. 15. Potential Energy or Wavelength Diagram for photon propagation along a galaxy-emitter to galaxy-observer path.

If the galaxy environment influences radiation redshifting, then microwave background radiation (CBR) is also affected and should display a wavelength offset equivalent to the ratio of expansion parameter values between present-day observer galaxy-space and intergalaxy-space;

$$CBR\ offset \equiv \left(\frac{a_{observer}(r, today)}{a_{space}(today)} \right) = \left(\frac{e^{\mu/2}(r, today)}{a(today)} \right) ;$$

$$Actual\ temp_{CBR} = Measured\ temp_{CBR} * CBR\ offset \ . \qquad (43)$$

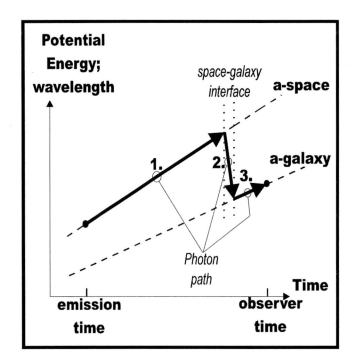

Fig. 16. Potential Energy or Wavelength Diagram for photon propagation along a microwave-emitter to galaxy-observer path.

A radiation evolution and propagation diagram is shown in Fig. 16 and, as illustrated, suggests an actual inter-galaxy background radiation temperature lower (longer wavelength) than measured inside the galaxy. However, since $a_{space}(today)$ is probably greater than that in the outer regions of the galaxy, a less-than-one CBR offset is predicted for observers in these regions. That is, appropriate galaxy models, with observers in high mass-density regions, would predict a smaller intergalaxy CBR temperature than that which is measured, a phenomenon that derives from the local space warping produced by matter in the vicinity of the detector. The outer regions of the cores, or hole-bulge masses, described here display a rapidly decreasing radial warping (see Fig. 4), however, and, when considered as emission sources, would apparently cause measurable impact on galactic spectrometric electromagnetic frequency shifts only when the emissions emanated from regions very close to the cores.

4. SUMMARY

In conclusion, we have modeled radiation generated galaxy cores, born in early time frames, exhibiting collapse zones with black hole type characteristics. The galaxy-cores, as modeled, display radius-dependent curvature, pressure and expansion rates and exhibit time evolution rates approaching the background $t^{1/2}$ dependencies in the outer regions while the inner regions are collapsing. We also postulate from this modeling that cosmological redshift data are interpretable as measurements of a localized galactic expansion parameter, both at the emitter and at the observer, and that cosmic microwave background radiation measurements should be impacted by the difference between galaxy and intergalaxy expansion rates.

REFERENCES

[1] Hubble, E. *The Realm of the Nebulae*; Yale: New Haven, CT, 1936.
[2] Einstein, A. in *Modern Cosmology in Retrospect*; Bertotti, B. et al; Cambridge University Press: Cambridge, U.K., 1990; p. 102.
[3] Silk, J. *The Big Bang*; W. H. Freeman: New York, NY, 1998.
[4] Ferris, T. *The Whole Shebang*; Simon and Schuster: New York, NY, 1997.
[5] Kormendy, J.; Richstone, D. *ARA&A* 1995, 33, 581.
[6] Magorrian, J.; Tremaine, S.; Richstone, D.; Bender, R.; Bower, G.; Dressler, A.; Faber, S.M.; Gebhardt, K.; Green, R.; Grillmair, C.; Kormendy, J.; Laver, T. *Astron. J* 1998, 115, 2285.
[7] Kormendy, J.; Gebhardt, K.; Richstone, D. *Bull. Am. Astron. Soc.* 2000, 32, 702.
[8] Gebhardt, K.; Bender, R.; Bower, G.; Dressler, A.; Faber, S. M.; Filippenko, A. V.; Green, R.; Grillmair, C.; Ho, L.; Kormendy, J.; Lauer, T. R.; Magorrian, J.; Pinkney, J.; Richstone, D.; Tremaine, S. *Astrophys. J. Lett.* 2000, 539, L13.
[9] Einstein, A. *The Meaning of Relativity*; Princeton: Princeton, NJ, 1955.
[10] Tolman, R. *Relativity, Thermodynamics and Cosmology*; Dover: Mineola, NY, 1987.
[11] Linder, E. *First Principles of Cosmology*; Addison Wesley: Essex, England, 1997.

INDEX

A

affect, 33, 74, 108
age, viii, 3, 27, 32, 33, 34, 85, 140, 144
alternative, 4, 15, 37, 50, 145
anisotropy, 47
annihilation, 99
Argentina, 49, 76
argument, 68, 89
ASI, 26
assumptions, 30, 83
asymptotics, 109, 110, 115

B

background radiation, 142, 148
baryon(s), 145, 146
behavior, ix, 7, 46, 47, 53, 59, 103, 104, 108, 110, 114, 115, 125, 126, 129, 130, 131, 136, 140, 142, 144, 145, 146, 147
Bianchi identity, 6
Big Bang, 149
birth, 126, 131, 132, 137, 138, 139, 140, 141, 145
black hole, ix, 32, 103, 104, 107, 108, 109, 117, 118, 119, 122, 123, 125, 126, 148
Boltzmann distribution, 146
BTZ black hole, 103, 104, 107, 119
burn, 28

C

candidates, 2, 3, 32
cast, 68, 83
catastrophes, 29
causality, 138
cellular automaton, 82, 92, 94
CERN, 33
classes, 47, 82, 83
classical mechanics, viii, 79, 81, 83, 100
classification, 113
closed string, 82
closure, 112
colonization, viii, 27, 32, 33, 34
communication, 34, 37, 38, 46, 47
community, 33
compatibility, 16
competition, 46
complement, 30
components, 4, 6, 104, 111
composition, 87, 111, 112, 116
comprehension, 72
concrete, 84
condensation, 2
configuration, 16, 52, 66, 67, 68, 70, 104, 106, 108, 109, 112
conflict, 81
conjecture, 30
conjugation, 71
conservation, 108, 112, 115, 117
construction, 40, 80, 81, 82, 85, 100, 107, 112, 114
context, 2, 4, 16, 32, 39, 46, 49, 57, 70, 80, 100, 104
continuity, 6, 7, 86
conversion, 90
correlation, ix, 16, 91, 125
correlation function, 91
cosmological space, 140
cosmological time, 138
coupling, 37, 52, 53, 66, 85, 99

D

dark energy, 4, 5, 7, 24, 32
dark matter, 32
death, 3
decomposition, 72
definition, 41, 42, 49, 58, 60, 66, 72, 74, 90, 91, 92, 112, 130
demand, 42, 54, 129
density, 5, 6, 39, 43, 104, 126, 127, 128, 129, 136, 137, 138, 139, 140, 141, 142, 144, 145, 146, 148
derivatives, 50, 55, 61, 94, 114, 115, 126
detection, 146, 147
differentiation, 6, 39, 128
diffusion, 28
discontinuity, 144
discretization, viii, 79, 83, 93, 94, 99
distribution, ix, 34, 85, 90, 91, 100, 101, 125, 126, 129, 138, 145
distribution function, 145
divergence, 115
DNA, 29
domain, vii, ix, 3, 37, 125, 144
dominance, 140, 141

E

early universe, ix, 3, 38, 49, 54, 125, 126, 140, 145
effective field theory, 52
elasticity, 38
election, 68
emergence, 80
emission, 146, 147, 148
energy density, 2, 4, 7, 39, 127, 130, 131, 138, 140, 141, 144, 146
energy momentum tensor, 44
entropy, 122, 136, 142, 144, 145
environment, ix, 125, 126, 130, 146, 147
equality, 115, 117
equilibrium, 81, 138
ethical standards, 30
ethics, 30
evolution, viii, 10, 13, 47, 49, 50, 53, 54, 55, 56, 58, 65, 66, 67, 68, 69, 70, 72, 74, 79, 80, 81, 82, 83, 85, 86, 87, 88, 89, 90, 91, 100, 101, 130, 131, 144, 145, 147, 148
expectation, 2, 83, 90, 91, 93
experts, 28
expression, 42, 45, 58, 62, 64, 85, 88, 91, 109, 112, 114, 117, 144, 146

F

family, 109, 110, 118
field theory, 51, 52, 81, 123
financial support, 47
first generation, 28
fluctuations, ix, 125, 126
fluid, vii, 1, 5, 6, 8, 39, 126, 128, 144, 145
freedom, 13, 41, 50, 53, 58, 65, 66, 67, 70, 72, 81, 82, 85, 94, 95, 99, 100, 101, 104

G

galaxy, viii, ix, 27, 28, 29, 30, 31, 32, 33, 34, 125, 126, 127, 138, 144, 145, 146, 147, 148, 149
gauge group, 112
gauge invariant, 81
gauge theory, 67, 103, 105, 122, 123
General Relativity, 1, 27, 32, 37, 47, 49, 52, 67, 69, 76, 79, 102, 103, 125
generalization, 43, 61, 104, 106, 109, 110
graph, 70
gravitation, vii, 52, 54, 68
gravitational constant, 2
gravitational field, 50, 106
gravitational lensing, 32
gravitational pull, 33
gravity, vii, viii, ix, 4, 24, 32, 37, 39, 50, 81, 102, 103, 104, 105, 106, 108, 119, 121, 122, 123

H

Hamiltonian, viii, 10, 12, 13, 49, 50, 53, 54, 55, 56, 57, 58, 59, 60, 61, 62, 63, 64, 65, 66, 67, 68, 69, 70, 71, 72, 73, 74, 75, 79, 80, 81, 82, 83, 84, 85, 86, 87, 88, 90, 91, 92, 93, 94, 95, 96, 97, 98, 99, 100, 104, 112, 113, 114, 117, 120, 123
Hilbert space, 56, 57, 80, 82, 86, 100
homogeneity, 54
hypothesis, 30

I

ideas, 33, 103
identification, 55, 65
identity, 97, 105, 106
inclusion, 39, 50
indices, 51, 83, 84, 104
infinite, vii, viii, 13, 15, 27, 30, 32, 33, 96, 99
inflation, viii, 5, 24, 27, 30, 32, 33, 39, 145, 146
influence, viii, 27, 32, 33, 93, 103, 104, 108, 115, 120

Index

input, 136
insertion, 88, 92
instability, 101
integration, 45, 50, 53, 62, 74, 87, 89, 91, 101, 114, 115, 117, 130, 144, 146
interaction(s), 66, 85
interest, 3, 8, 29, 32, 37, 38, 40, 72
interface, 146, 147
interpretation, 5, 16, 37, 52, 55, 57, 67, 68, 72, 138
interval, 90, 138, 139, 147
inversion, 69

K

knowledge, 4

L

Lagrange multipliers, 53
laws, vii, viii, 27, 33, 34, 105, 108, 112, 114, 115, 117
lead, 54, 82, 83, 85, 96, 109, 121, 145
lifetime, 28
limitation, 24
localization, 95, 99
locus, 16, 19, 20, 21

M

manifolds, 3
mapping, 82, 85
mass, ix, 2, 39, 80, 83, 92, 94, 125, 126, 133, 135, 136, 137, 138, 139, 140, 141, 142, 144, 145, 146, 148
matrix, 80, 83, 87, 88, 91, 99
memory, 85, 101
methodology, viii, 37, 38, 39, 43, 45
Milky Way, 34, 126
minisuperspace, viii, 13, 15, 16, 49, 50, 53, 54, 65, 72
mixing, 32
modeling, 28, 128, 145, 148
models, vii, viii, 1, 2, 3, 4, 8, 27, 30, 32, 33, 38, 46, 49, 50, 51, 53, 54, 55, 56, 57, 58, 65, 67, 72, 73, 79, 80, 81, 82, 83, 85, 92, 99, 100, 101, 103, 148
modulus, 84
momentum, vii, 1, 4, 6, 16, 33, 53, 55, 56, 60, 66, 67, 68, 69, 74, 80, 91, 93, 97, 106, 107, 114, 115, 116, 118, 119, 121, 126
motion, 8, 11, 44, 46, 52, 66, 69, 70, 71, 72, 73, 80, 81, 82, 83, 84, 85, 87, 88, 89, 91, 93, 94, 99, 100, 113, 126

motivation, 46, 81, 83, 84
movement, 69
multidimensional, 33
multiplier, 66, 68, 84, 93

N

NATO, 26
natural evolution, 30
nature of time, 54
needs, 4, 16, 85, 89, 99
neutrinos, 32
normalization constant, 15

O

observations, vii, 1, 4, 6, 8, 10, 12, 24, 81
obstruction, 50, 65
operator, 12, 56, 58, 60, 64, 71, 72, 74, 83, 87, 88, 89, 90, 91, 92, 93, 94, 95, 97, 99, 100
ordinary differential equations, 47
organic matter, 28

P

parameter, vii, viii, 1, 2, 3, 6, 10, 11, 12, 17, 23, 39, 50, 51, 54, 55, 66, 72, 79, 80, 82, 83, 93, 100, 127, 131, 136, 145, 146, 147, 148
partial differential equations, 99
particle creation, 146
particle physics, 2, 3, 4
permit, 57
perspective, 32, 81, 82, 144
physics, vii, viii, 16, 24, 27, 33, 34, 38, 80, 145
planets, vii, 30, 33
pleasure, 101
power, 4, 42, 46, 47, 51, 146
prediction, 16, 30
present value, 2, 3, 7, 12
pressure, vii, 1, 4, 5, 6, 126, 127, 128, 129, 130, 136, 142, 144, 145, 148
principle, vii, 5, 8, 9, 13, 24, 38, 40, 52, 59, 81, 92, 114, 115, 120
probability, viii, 30, 49, 55, 57, 79, 82, 85, 86, 90, 100
probability distribution, viii, 79, 82, 86, 90
program, 50
propagation, 131, 136, 146, 147, 148
propagators, 57

Q

quantization, viii, 12, 13, 49, 50, 53, 55, 56, 57, 58, 59, 61, 63, 64, 72, 74, 82, 83, 95, 101, 103
quantum cosmology, vii, viii, 1, 2, 3, 5, 12, 13, 16, 24, 50, 66, 72
quantum field theory, 2, 38, 50, 102
quantum fields, 2
quantum gravity, 16, 24, 55, 80, 81, 102, 103
quantum mechanics, 16, 50, 55, 69, 82, 92, 95, 100
quantum state, 61, 63, 71, 72, 83
quantum structure, 119
quantum theory, 16, 56, 79, 80, 90

R

radiation, ix, 3, 125, 126, 127, 129, 130, 131, 133, 135, 136, 137, 138, 139, 140, 141, 142, 144, 145, 146, 147, 148
radius, 3, 4, 5, 127, 128, 129, 131, 132, 135, 136, 137, 138, 139, 144, 145
range, 6, 96, 100, 126, 143
reading, 82, 119
real time, 12, 67
reasoning, 30, 35
recall, 39, 65, 74, 84, 116
recalling, 53
recovery, 147
redshift, 146, 148
reduction, 38, 39, 44, 47, 50, 56, 61, 64, 67, 138, 144, 145, 146
redundancy, 86
reference frame, 39
reference system, 68
relationship, 126, 142, 145
relativity, vii, ix, 8, 24, 80, 81, 92, 103, 104, 123
relevance, viii
resolution, 82
rotations, 105

S

scalar field, viii, 2, 4, 37, 38, 39, 40, 43, 81
scaling, 74
Schwarzschild solution, 132
search, 33, 72
second generation, 28
security, 30
selecting, 51, 73, 100
self, 56, 58, 145
separation, 37, 50, 53, 54, 85
series, 34, 79
shares, 89
sign, 57, 71, 99
signals, 28, 30
Singapore, 25, 76, 77, 121
spacetime, vii, viii, 1, 3, 4, 5, 6, 7, 16, 27, 32, 50, 51, 52, 53, 54, 75, 104, 105, 106, 107, 108, 112, 115, 119
special relativity, vii
special theory of relativity, vii
species, 30, 139, 140
spectrum, 51, 80, 90, 92, 93, 94, 95, 96, 99, 145
speed, 66
spin, 106
stability, 32
stages, 72
Standard Model, 80
standards, 30
stars, 28, 30, 32, 33
statistics, 101
stochastic quantization, 95
strength, 51, 54
stress, 95
stretching, 147
string theory, viii, 3, 49, 50, 51, 52, 58, 80, 92, 118
substitution, 39, 71, 108, 113, 129
summer, 27
Sun, 28, 30, 32
supergravity, 4
supernovae, 28
supersymmetry, 32
SUSY, 2
symmetry, viii, 2, 3, 12, 15, 49, 50, 51, 67, 69, 70, 71, 73, 81, 83, 86, 90, 101, 102, 103, 104, 108, 110, 112, 115, 117, 118
systems, viii, 47, 50, 61, 62, 66, 67, 68, 70, 73, 79, 80, 81, 82, 92, 99, 100, 121, 123

T

technology, 33, 38
temperature, ix, 125, 135, 136, 140, 141, 144, 145, 146, 148
tension, 51, 52
theory, vii, 4, 24, 50, 51, 52, 54, 56, 66, 67, 68, 71, 72, 79, 80, 83, 95, 103, 104, 105, 106, 118, 122, 123
Theory of Everything, 33
time, vii, viii, ix, 2, 3, 4, 5, 7, 8, 10, 12, 27, 28, 30, 38, 39, 40, 45, 46, 49, 50, 51, 54, 55, 56, 57, 58, 59, 60, 61, 63, 64, 65, 66, 67, 68, 69, 70, 71, 72, 73, 74, 75, 79, 80, 81, 82, 83, 84, 85, 86, 87, 88, 89, 90, 91, 92, 93, 94, 95, 96, 100, 101, 114, 116,

125, 126, 127, 128, 129, 130, 131, 138, 139, 140, 141, 142, 144, 145, 146, 147, 148
time increment, 138
time pressure, ix, 125, 126
time variables, 56, 57, 87
topology, 5, 12, 56
total energy, 5, 66, 126, 140, 144
trajectory, 68, 69, 70, 71, 73, 81, 84, 86, 100
transformation, 10, 11, 12, 51, 55, 56, 58, 59, 61, 62, 69, 70, 71, 72, 73, 74, 75, 86, 95, 105, 107, 110, 112, 114, 117
transformations, 51, 61, 69, 70, 72, 80, 99, 100, 109, 110, 111, 112, 114
transition, 132, 146
translation, 89, 114, 116
transport, 105
triggers, 81
tunneling, vii, 1, 32, 100, 101

U

uncertainty, 138
universe, vii, 1, 2, 3, 4, 5, 6, 7, 8, 10, 11, 12, 13, 14, 15, 16, 23, 24, 39, 49, 53, 54, 58, 59, 63, 66, 72, 82, 90, 100, 126, 138, 140, 146

V

vacuum, 2, 4, 7, 97, 104, 106, 107, 108, 109, 110, 119, 123

validity, 74
values, ix, 8, 10, 11, 12, 15, 17, 38, 39, 56, 58, 73, 83, 85, 86, 89, 91, 92, 93, 103, 104, 112, 117, 119, 132, 134, 136, 137, 140, 141, 147
variable(s), 2, 8, 13, 14, 38, 45, 46, 55, 56, 57, 58, 59, 60, 61, 62, 63, 64, 66, 67, 68, 69, 70, 71, 72, 74, 75, 81, 84, 87, 89, 90, 95, 101, 105, 108, 112, 114, 115, 131, 145
variation, 12, 66, 115
vector, 51, 73, 94, 95
vein, 95
velocity, 47, 126, 131, 133, 136, 142

W

weakness, 32
Wheeler-DeWitt equation, vii, 1, 5, 13, 14, 57, 60, 61, 62, 63, 64, 65, 81
work, vii, 5, 16, 24, 32, 38, 44, 46, 47, 72, 80, 83, 87, 93, 101, 119, 125, 126

Y

Yang-Mills, 122
yield, vii, 1, 2, 4, 44, 46, 75, 80, 99, 113, 119, 126